The Flavobacterium-Cytophaga Group

GBF



Gesellschaft für
Biotechnologische
Forschung mbH
Braunschweig-Stöckheim

Edited by H. Reichenbach and O.B. Weeks

Proceedings of the International Symposium
on Yellow-Pigmented Gram-Negative Bacteria
of the Flavobacterium-Cytophaga Group
8 to 11 July 1980



# The Flavobacterium-Cytophaga Group

Verlag Chemie · Weinheim · Deerfield Beach, Florida · Basel · 1981

GBF Monograph Series, No. 5 (March, 1981)

CIP-Kurztitelaufnahme der Deutschen Bibliothek

**International Symposium on Yellow-Pigmented Gram-Negative Bacteria of the Flavobacterium-Cytophaga ⟨1980, Braunschweig⟩:**
Symposium on the Flavobacterium-Cytophaga Group: proceedings of the Internat. Symposium on Yellow-Pigmented Gram-Negative Bacteria of the Flavobacterium-Cytophaga Group, July 8 to 11, 1980 / GBF, Ges. für Biotechnolog. Forschung mbH Braunschweig-Stöckheim. Ed. by H. Reichenbach and O. B. Weeks. – Weinheim; Deerfiedl Beach, Florida; Basel; Verlag Chemie, 1981.
(GBF monograph series; No. 5)
ISBN 3-527-25919-8
NE: Reichenbach, Hans [Hrsg.]; Gesellschaft für Biotechnologische Forschung ⟨Braunschweig⟩: Schriftenreihe der GBF; HST

Edited by Prof. Dr. H. Reichenbach, Gesellschaft für Biotechnologische Forschung, Dept. of Microbiology, Braunschweig, and Prof. Dr. O. B. Weeks, Arts and Sciences Research Center, New Mexico State University, Las Cruces, USA

Copy Editor Dr. J.-H. Walsdorff, Gesellschaft für Biotechnologische Forschung, Braunschweig



**Graphik-Design Team Zisowsky + Oehring, Braunschweig**

In Memoriam
Professor Dr. Owen B. Weeks
1916 - 1981

Dr. O.B. Weeks suddenly died January 8, 1981, when the compilation of this proceedings volume was in its final stage. Dr. Weeks was an authority in the field of flavobacterial taxonomy. He was responsible for the section on the genus Flavobacterium in Bergey's Manual of Determinative Bacteriology, 8th edition, and he served for a number of years on the International Committee on Systematic Bacteriology, and as chairman of the Advisory Committee on Gram-Negative Aerobic Rods of the Bergey's Manual Trust. Dr. Weeks' research included, however, also lipid metabolism in bacteria and bacterial carotenoids. He was host and organizer of the Second International Congress on Carotenoids in 1969 at New Mexico State University in Las Cruces. Dr. Weeks' scientific work is documented in more than 50 publications. The participants of this meeting will keep him in memory as the amiable and scholarly co-organizer of the First International Symposium of the Flavobacterium-Cytophaga Group.

# FOREWORD

The two bacterial genera, Cytophaga and Flavobacterium, account for many of the yellow or orange pigmented, gram-negative bacteria which we isolate from natural and man-made environments. However, their taxonomy is far from being clear. This is very unsatisfactory, particularly as many of these bacteria are of considerable practical importance, appearing quite regularly in the clinical laboratory, or playing a role as spoilage organisms in the food industry, as fish-pathogens, or as producers of interesting enzymes and antibiotics. Last but not least, enormous populations of these organisms live in soil, freshwater, and marine environments, and are undoubtedly greatly involved in the degradation and turnover of organic substrates in nature, as well as in sewage plants.

Both "genera" comprise rod-shaped, heterotrophic, primarily aerobic bacteria, which, in the case of the cytophagas move by gliding, while the flavobacteria are either immotile or flagellated. As a result of the research of the past 15 years we can today be sure that both groups are taxonomically heterogeneous. Somewhat unexpectedly, observations came to the light which suggested connections, perhaps even relatively close ones, between some of the flavobacteria with a low DNA base ratio (30 to 40 mol % G+C) and respective Cytophaga-like bacteria (CLB). This meeting has been organized by us, together with Dr. D. Claus from the DSM (German Collection of Microorganisms) in Göttingen to discuss these new developments. Some of the more impressive data which would indicate a relationship between flavobacteria and CLB are presented in this volume, e.g. studies on the distribution among these organisms of lipoquinones, hydroxy fatty acids, and flexirubin-type pigments. Reports on unusual cellular constituents, like sphingolipids or sulfonolipids, in some of these bacteria should stimulate investigation into the occurrence of such compounds in other representatives of this group and might provide valuable additional indications of a relationship.

Of course, further problems still remain to be solved. Thus, it is not yet apparent how the new genera, Cytophaga and Flavobacterium, should be delimitated and defined, nor how to proceed with the remaining organisms, e.g. the "flavobacteria" with a high G+C content (60 to 70 mol %). As can be seen from the contributions of this symposium, lively discussions take place about such questions, but there is neither universal agreement yet, nor are there always sufficient available data to enable the taxonomic problems to be dealt with competently. There are actually large areas in the taxonomy of flavobacteria and CLB, where we are still only at the start. Modern taxonomic methods, like those offered by molecular biology, or the chemosystematic approach, as applied more recently to this difficult group of bacteria, promise progress in the near future. It is our opinion that this symposium has provided a good survey of the state of the taxonomy of flavobacteria and CLB, and we hope that it will stimulate further research.

We should like to take the opportunity to thank all those persons who helped to make this symposium possible. The directors of the GBF, Prof. Dr. K. Kieslich and Dr. H. Zeitträger, were in favor of our enterprise from the very beginning, and allowed us to use the lecture room as well as the facilities of the institute. They also gave us essential financial support. R. Radloff solved most of the organizational problems, and Karin Rahn typed many letters in connection with the symposium and had much work with the articles for this proceedings volume. Dr. J.H. Walsdorff gave valuable assistance in compiling and editing this volume.

Las Cruces, December 1980
Braunschweig, March 1981

The Editors

## CONTENTS

LIST OF AUTHORS

# A SURVEY OF *FLAVOBACTERIUM*

O.B. Weeks

Arts and Sciences Research Center
New Mexico State University
Las Cruces, New Mexico 88003 USA

## Summary

The survey of *Flavobacterium* is a general review of the origins of the genus and of its taxonomic development in the various editions of Bergey's Manual. Brief consideration is given to the chemistry of the pigments which have been studied.

The title, A Survey of *Flavobacterium*, is a generalization that will allow me to speak broadly on the subject. I have planned three topics: the beginning of the genus; the development of the artificially conceived color genus primarily in the successive editions of Bergey's Manual; and finally some thoughts about the future of *Flavobacterium*. The survey will not include recent taxonomic proposals since these will be presented by Barry Holmes later in the program.

Most persons who have had a taxonomic interest in *Flavobacterium* know that the genus was proposed first by the S.A.B. Committee on bacterial classification. The proposal in its entirety was:

> "*Flavobacterium* *gen. nov.* Rods of medium size, occurring in water and soil, forming a yellow to orange pigment on culture media. Characterized by feeble powers of attacking carbohydrates, occasionally forming acid from hexoses but no gas. Motile or non-motile. Generally Gram-

negative. The type species is Flavobacterium aquatile (Frankland)" [4, p. 97].

This is the only published description of the new genus and I have presented it without the explanations which might help understanding. The S.A.B. Committee which arranged the first Bergey's Manual, had for its members D.H. Bergey, F.G. Harrison, R.S. Breed, B.W. Hammer and F.M. Huntoon. Dr. Bergey was chairman. The emphasis in the manual was upon determinative aspects of bacterial taxonomy. The overall arrangements followed recommendations of R.E. Buchanan who had presented the details in a series of publications on bacterial classification and nomenclature [6]. Breed and Buchanan assumed major roles in preparation of the succeeding editions of Bergey's Manual. Probably the predecessor to the Manual was Chester's Manual of Determinative Bacteriology [9] and his earlier publication in the Annual Report of Delaware College Experiment Station [8]. Both were tabulations of previously described bacteria arranged in form of differential keys. The S.A.B. Committee arranged their classification as proposed previously by Buchanan. The order Eubacteriales, one of six proposed for Schizomycetes, was divided into five families and one of these, Bacillaceae, was further subdivided into eleven tribes. Two of these are relevant, Chromobactereae and Achromobactereae. The tribe for pigmented bacteria included four genera one of which was Flavobacterium. All of the color-based genera have survived but not the colorless counterpart, Achromobacter. Flavobacterium contained forty-six species most of which had previously been named Bacillus and a few Bacterium, Pseudomonas or Vibrio. There is no evidence to decide how many of the species were represented by cultures but probably only a few. It seems probable that many of the species existed as descriptions only. This is the reason I have said previously that Flavobacterium is more an historic concept than a taxonomic reality [16]. Breed very soon became the person responsible for taxonomic arrangements of Flavobacterium in successive editions of Bergey's Manual. I believe Bergey prepared the initial arrangement of the genus. The general taxonomic problems confronting Breed were the topic of his presidential address to the Society of American Bacteriologists, 1927 [5]. Breed was very much dedicated to Bergey's Manual and Flavobacterium was one of his special interests.

The second of my topics is taxonomic development of Flavobacterium. The S.A.B. Committee very much believed in the type concept and one of Breed's goals was to obtain a culture to represent Flavobacterium aquatile which Bergey had arbitrarily chosen as the type species. Another of his projects was to remove from the genus all the species which he thought were unidentifiable. He persisted with these two projects until his death in 1956.

The search for a culture to represent F. aquatile apparently commenced very early. Bergey in his younger days worked with Drs. Wright, Ravenel and Abbott who made extensive bacteriological studies of the tap water of Philadelphia between 1890 and 1900. The impetus was an outbreak of cholera in Philadelphia in 1890. Apparently

Bergey used some of the cultures that were isolated in his teaching at the University of Pennsylvania. Dr. Breed sent me this information in a letter dated December 20, 1954. He said also, "In other words, I think you can put down Dr. Bergey's description of *F. aquatile* as given in the first edition of the Manual as another case of wrongful identification of a yellow organism isolated from water" [personal communication]. This may explain the characterization of *F. aquatile* Frankland as a peritrichous flagellate in the first edition of Bergey's Manual.

Breed commenced a search for a culture to represent *F. aquatile* in about 1940 when he asked E. Windle Taylor, Metropolitan Water Board, London, to isolate pigmented bacteria from some of the deep wells that supplied water for London. It was Breed's idea that a return to the habitat from which Percy and Grace Frankland initially isolated their *Bacillus aquatilis* [11] might produce a similar bacterium. Identification of their species would have been an act of great faith and nothing much more substantial. Taylor seems to have made quite a few isolations commencing about 1940 and continuing upon request until about 1952. I believe the last culture which he sent was that now designated *F. aquatile* strain Taylor ATCC 11947. One of the earlier isolates sent to Grant Stahly, Ohio State University, was distributed fairly widely and eventually became *F. aquatile* ATCC 8375. A detailed account of the cultures sent to the United States has already been given [15] and is of little importance now with two exceptions. *F. aquatile* ATCC 8375 is coryneform in nature and apparently a variety of *F. dehydrogenans* which I have already renamed *Cellulomonas dehydrogenans* [18]. The other matter is that Taylor was never asked to make cultural studies of the yellow chromogens that he isolated from the deep wells in the chalk areas of England. Breed also made no studies of any of the cultures which usually were designated *F. aquatile*. *F. aquatile* ATCC 11947 now represents the species but it does not have enthusiastic endorsement for this position.

The second major objective of Breed was to remove species that to him seemed unidentifiable. He persisted with the task and at the publication of Bergey's Manual, 6 edition, placed such species into an appendix of the genus. In a letter dated January 7, 1955, he complained that, ". . .everybody dumped poorly known species into *Achromobacter* and *Flavobacterium* until they became dumping grounds" [personal communication]. This has been said quite a number of times by more persons than Dr. Breed. The removal of unwanted species from *Flavobacterium* has been a continuous action almost from the start of the genus. The problem was not simply a removal of vaguely defined species from the original collection, it was also a matter of removing Gram-positive species, newly described species which had been poorly characterized, if at all, and species for which no representative cultures existed. The most recent edition of Bergey's Manual showed *Flavobacterium* to have 12 species and an extensive appendix of unsuitable species [17]. The suitability of the reference cultures of three of the twelve species has since been questioned. These are: *F. aquatile* ATCC 11947, *F. devorans* ATCC 10829 and *F. rigense* strain F18. The latter two apparently have polar flagella [Dr. E. Yabuuchi and Dr. M. Hendrie, personal communications]. The F. aquatile culture has been said to show gliding movement [13].

F. aquatile ATCC 11947 probably represents the last attempt to legitimize the present type species of Flavobacterium. It is difficult to justify retaining the culture in this capacity except for arbitrary reasons. F. aquatile persisted for many years because Bergey and Breed decreed it should. Later it persisted because I was convinced that given the original description, or perhaps lack of description, if a culture could be found which had the general attributes of Flavobacterium and F. aquatile, it might be acceptable as a satisfactory reference culture. The problems are knowing the general attributes and finding the reference culture, neither of which has proven an easy task. When I chose F. aquatile ATCC 11947 [15] taxometric methods were scarcely started, DNA nucleotide base ratio analysis was not yet a concept, chemotaxonomy was not applied to bacterial, the first bacterial pigment had not yet been characterized precisely and I was only vaguely aware of the phenetic similarity between some cytophagas and flavobacteria.

Taxonomic irresponsibility evident in the improperly described species reduced the stature of Flavobacterium and lessened its taxonomic reality. While this is not unusual in bacterial genera, Flavobacterium has been especially attractive because color is a conspicuous attribute and the genus was founded to foster this attraction. Irresponsibility in assignment of species to Flavobacterium has not been limited to the early development of the genus, it continues to the present. Often it is not the intention of an investigator to be taxonomically irresponsible. An investigator who finds his organism in an unusual environment, e.g., F. oceanosedimentum [7], or with unusual physiological qualities, e.g., Flavobacterium sp. 0147 [12] or Flavobacterium sp. CIP 12-75 [14], is almost compelled to make a taxonomic assignment. Usually such investigators do not wish to do taxonomic work. Irrespective of the intention, however, taxonomic indiscretions should be discouraged if the genus is to be more than the collecting heap it has become [10].

The last topic, pigmentation, does not yet have the data required for a detailed discussion. Studies of the chemistry of flavobacterial pigments has been a neglected subject. The best studied culture is Flavobacterium sp. 0147 (R1519) [12]. This marine bacterium is a highly pigmented culture which was selected from among the progeny of an X-radiated parent population by investigators at Hoffman-LaRoche, Basle. The objective was to obtain a microbial source of an oxygenated carotenoid. The culture contains only carotenoids and the principal compound is zeaxanthin. F. aquatile ATCC 11947 also contains a β-carotene, zeaxanthin carotenoid system [O.B. Weeks, unpublished study]. I know of no other studies of flavobacterial pigments except the casual work in which tentative characterization was based upon electron absorption spectrometry. This is not satisfactory since spectral absorption is due to the chromophore and not necessarily unique to a single class of compounds. My interest is the chemistry of bacterial pigments but somewhat ironically, I have worked primarily with the decapreno carotenoid systems of F. dehydrogenans, a Gram-positive coryneform bacterium which I have renamed Cellulomonas dehydrogenans [18]. Very recently Dr. Reichenbach and his colleagues have characterized the pigments of a typical flavobacterial culture obtained from Pay Hayes [2]. I am fascinated by their finding that the pigments are flexirubins, those unique compounds which these investigators have associated with cytophagas [1]. About two months ago I commenced work with two cultures representing F. breve and F. odoratum. The principal pigments of these cultures are not carotenoids and their general properties suggest they are aryl polyene esters.

The taxonomic and physiologic studies reported since about 1970 show Flavobacterium to have considerable stature. Data of taxometric analyses, of molecular biological studies, of chemotaxonomic investigations seem to justify the existence of Flavobacterium. The interrelationships of non-motile, chromogenic bacteria, however, are a problem which may not be resolved easily. The nature of the flavobacterium-cytophaga complex [10] is not yet very apparent but this is not new.

## References


[1] Achenbach, H., Kohl, W., Reichenbach, H.: Revista Latinoamericana de Quimica: 9, 111-124 (1978)

[2] Achenbach, H., Kohl, W., Böttger-Vetter, A., Reichenbach, H.: Tetrahedron (1981, in print)

[3] Andrewes, A.G., Hertzberg, S., Liaaen-Jensen, S., Starr, M.P.: Acta. Chem. Scand. 27, 2383-2395 (1973)

[4] Bergey, D.H., Harrison, F.C., Breed, R.S., Hammer, B.W., Huntoon, F.M.: Bergey's manual of determinative bacteriology. Baltimore (1923): The Williams and Wilkins Co.

[5] Breed, R.S.: J. Bact. 15, 143-163 (1928)

[6] Buchanan, R.E.: J. Bact. 1, 591-596 (1916); ibid. 2, 155-164 (1917); ibid. 2, 347-350 (1917); ibid. 2, 603-617 (1917); ibid. 3, 27-61 (1918); ibid. 3, 175-181 (1918); ibid. 3, 301-306 (1918); ibid. 3, 403-406 (1918); ibid. 3, 461-474 (1918); ibid. 3, 541-545 (1918)

[7] Carty, C.E., Litchfield, C.D.: Int. J. Syst. Bact. 28, 561-566 (1978)

[8] Chester, F.D.: p. 53-145 in: Ninth annual report. Delaware College Agr. Exp. Sta. Wilmington (1897): Mercantile Printing Co.

[9] Chester, F.D.: A manual of determinative bacteriology. London (1909): Macmillan Co.

[10] Christensen, P.: Can. J. Microbiol. 23, 1599-1653 (1977)

[11] Frankland, G.C., Frankland, P.F.: Z. Hyg. Infektionskrankh. 6, 373-400 (1889)

[12] McDermott, J.C.B., Britton, G., Goodwin, T.W.: J. gen. Microbiol. 77, 161-171 (1973)

[13] Perry, L.B.: J. appl. Bact. 36, 227-232 (1973)

[14] Pichinoty, F., Bigliardi-Rouvier, J., Mandel, M., Greenway, B., Méténier, G., Garcia, J.L.: Antonie van Leeuwenhoek 42, 349-354 (1976)

[15] Weeks, O.B.: J. Bact. 69, 649-658 (1955)

[16] Weeks, O.B.: J. appl. Bact. 32, 13-18 (1969)

[17] Weeks, O.B.: p. 357-365 in: Buchanan, R.E., Gibbons, N.A. (ed.), Bergey's manual of determinative bacteriology, 8 ed. Baltimore (1974): The Williams and Wilkins Co.

[18] Weeks, O.B., Francesconi, M.D.: J. Bact. 136, 614-624 (1978)

# THE CLASSIFICATION OF THE *CYTOPHAGA*-LIKE BACTERIA

H. Reichenbach, H. Behrens, I. Hirsch

Gesellschaft für Biotechnologische Forschung,
Mascheroder Weg 1,
D-3300 Braunschweig-Stöckheim, FRG

## Summary

Based on a thorough study of a considerable number of strains, most of them newly isolated, we believe that the taxonomy of the cytophagas and related organisms has to be reconstructed from its very base. Using GC and DNA-DNA hybridization data as well as chemosystematic, physiological, biochemical and morphological observations, the following groups can be recognized with some confidence: *Sporocytophaga*, *Flexibacter* (in the original sense), *Lysobacter*, and *Sphaerocytophaga*/*Capnocytophaga*. The remaining bulk of strains, classified now as *Cytophaga*/*Flexibacter* (the latter in a sense deviating from the original definition) comprises with certainty several different genera, and the genus *Cytophaga* *s. str.* may finally have to be restricted again to aerobic cellulose-decomposing soil organisms. With the exception of a few rare cases, our present knowledge does not allow us to reliably distinguish between species.

## Introduction

It is hardly exaggerated to say that at present the taxonomy of the gliding bacteria is in a desolate state. Not that it had been in a better state before, but now we know enough about many of these organisms to be reasonably sure that our current classification cannot be correct. This is particularly true for the large, diverse, and still relatively little studied assembly of organisms found in the latest edition of Bergey's Manual of Determinative Bacteriology under the heading Cytophagales.

There is little doubt that "The Gliding Bacteria" are phylogenetically heterogeneous. There are at least 4 different groups of gliding organisms which have little or nothing in common beyond their gram-negative cell wall and their way of movement.

a) The genus *Chloroflexus* comprises filamentous, facultatively phototrophic bacteria, which are obviously related to the chlorobacteria. b) There is a large group of gliding cyanobacteria. Again, most of these organisms are filamentous, but otherwise they are considerably diversified, morphologically as well as physiologically. Many of them are apochlorotic and depend on the utilization of organic substrates, sometimes in combination with an oxidation of sulfur or sulfur compounds (e.g. *Beggiatoa*, *Vitreoscilla*, *Herpetosiphon*, *Leucothrix*, *Saprospira*, *Simonsiella*). c) The largest group of gliding bacteria are heterotrophic unicellular rods, many of them with a remarkable morphogenetical potential. This is the group of the *Cytophaga*-like bacteria and myxobacteria, and both could be related closely enough to be united in one common class, Flexibacteriae /62/. d) Parenthetically it may be mentioned, that a kind of gliding motility is observed also with certain mycoplasmas /8/.

The groups mentioned above have not always been kept separate in earlier taxonomic schemes, and therefore a definition of what we mean by "*Cytophaga*-like bacteria" may be in place.

## Definition

The "*Cytophaga*-like bacteria" (CLB) are heterotrophic gram-negative unicellular rods that move by gliding and do not form fruiting bodies.

The latter restriction sets the CLB apart from the equally unicellular and gliding myxobacteria. It seems that these two groups of organisms can also be reliably distinguished by the GC-content of their DNA, which in the myxobacteria is always above 65 %, in the CLB usually between 30 and 50 %, although the lysobacters approach the lower limits of the GC-range of the myxobacteria.

Gliding motility is the main distinguishing character between the CLB and immotile flavobacteria with a low GC-content.

## Gliding Motility

As gliding motility is obviously of central importance in all classification schemes, a brief characterization of the phenomenon may be useful. The ability of an organism to move by gliding can be deduced from the following observations: a) The bacterium may form spreading colonies, or swarms (Fig.1). Spreading becomes apparent only on special substrates. We often find useful: yeast-agar (vy/2-agar: baker's yeast, 0.5 % by fresh weight of yeast cake; $CaCl_2$, 0.1 %; pH 7.2; autoclaved), and plain water-agar (agar, 1.5 %; $CaCl_2$, 0.1 %; pH 7.2) with or without a streak of living or dead *Escherichia coli*. There are, however, gliding bacteria that do not form spreading colonies under any known condition.

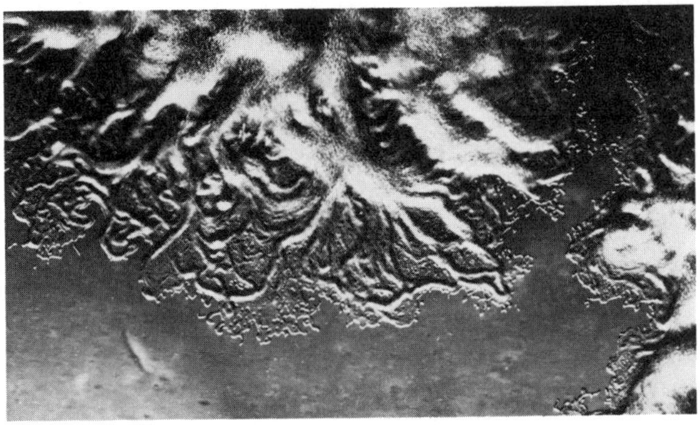

Fig. 1. Spreading colony of marine Cytophaga-like bacterium Ro 520. X 9.

b) Gliding movements can often be observed directly under the microscope. So far no organism is known that is sometimes flagellated and sometimes moves by gliding, so that all actively swimming bacteria must be regarded as not gliding. c) Occasionally gliding movements are so slow, that they cannot be recognized in the slide mount. In this case it may be helpful to start a micro-culture on a cover glass coated with a very thin agar layer and mounted on a slide in a wet chamber. In such cultures slime tracks can often be seen under the microscope when applying phase contrast (Fig. 2), or at least the arrangement of the cells in the micro-colony, and particularly at its edge, strongly suggests cell translocation by gliding (Fig. 3).

Of course, it is conceivable that the ability to glide could get lost by mutation, in which case classification of the respective strain would become impossible on the

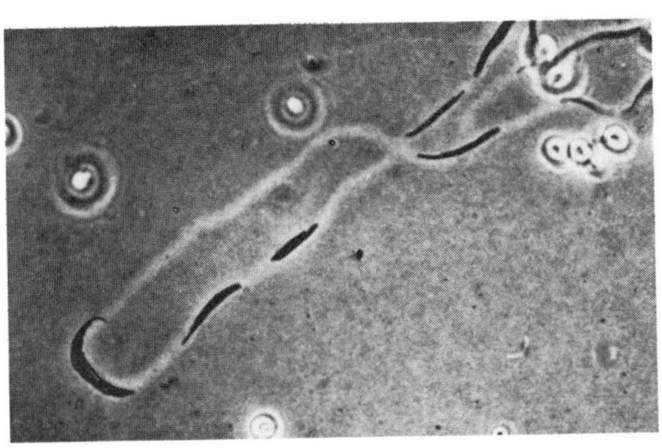

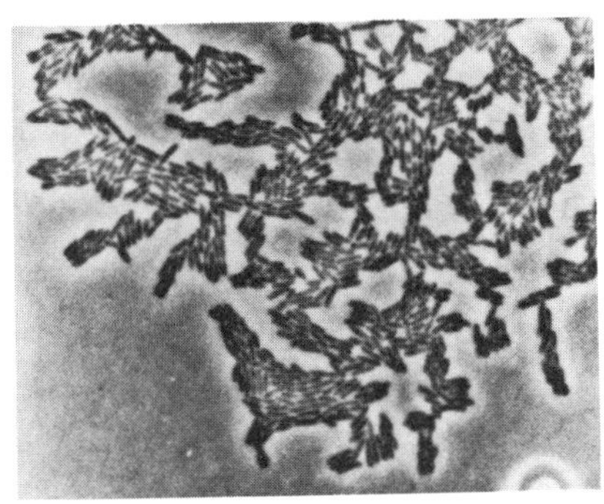

Fig. 2. Slime tracks on an agar surface. Cytophaga-like bacterium Cy Mb 632 in a chamber culture on yeast agar (the white dots are yeast cells). Phase contrast, X 475.

Fig. 3. Microcolony of Cytophaga-like bacterium Cy 637 in a chamber culture. The arrangement of the cells suggests gliding movements. Phase contrast, X 925.

basis of current definitions. It is not clear at the moment whether this is a serious problem. Perhaps certain flavobacteria with a low GC-content could be regarded as immotile CLB.

Finally it should be emphasized that, while there are several interesting hypotheses, the mechanism of gliding motility in CLB is not yet really understood /46, 47/, that there are organisms that behave unorthodoxically, and that it is not always easy to demonstrate gliding unequivocally /29/.

## The Organisms

Prior to discussing any taxonomic problems, a survey of the "raw material", the different organisms belonging to the Cytophaga-like bacteria (CLB) must be presented.

a) The first CLB that was described exactly enough to be reliably recognizable was the aerobic cellulose-decomposer Spirochaeta cytophaga in 1919 /33/. It was later shown to be a gliding bacterium and to form microcysts, i.e. spherical, desiccation-resistant resting cells /37/. It was finally renamed Sporocytophaga myxococcoides /63/. In 1929 the cellulose-decomposing true cytophagas were discovered in soil. In fact, the genus Cytophaga has been created for these organisms /68/. Much later, around 1950, cellulose-decomposing CLB were also found in marine environments; they reputedly did much damage to fishing gear at that time /36/. Particularly in the thirties and forties, the cellulose-decomposing CLB excited much interest, they were repeatedly studied in considerable detail, including their ecological and agricultural role, and there are several voluminous reports about this research /10, 19, 34, 36, 43, 45/.

b) The second group of CLB to be discovered were the fish-pathogens in 1922 /15/. Also in this case, the connection to the gliding bacteria became apparent only much later. The organism in question was first named Bacillus columnaris. It was later reclassified as a true myxobacterium and placed into the genus Chondrococcus /44/. Already in 1945 it was realized that this organism was actually a CLB /21/. It took, however, another 30 years before this fact was generally accepted. In the course of the years, additional fish-pathogenic CLB have been described, e.g. Cytophaga psychrophila, and marine fish-pathogens /1, 7/. For a long time fish-pathogenic gliding bacteria were known only in the U.S., but beginning in 1965 an increasing number of cases were reported from Japan and Europe. As these organisms may lead to severe economic losses, they were carefully studied at several places, and a number of summarizing research reports have been published /1, 5, 7, 13, 14/. The situation is, however, still far from being clear, e.g. nothing is known as to how many different fish-pathogens exist, nor whether and how they are connected with ordinary free-living strains, nor where they come from, nor to what extent the etiology of the diverse fish diseases can be attributed to these organisms.

c) The next group of CLB that came to light were the chitin-decomposers /35/. While they were recognized immediately as gliding bacteria, their connection with the CLB was realized only many years later /65/. The ability to degrade chitin is found in many strains of clearly different CLB, so that this character is of limited value taxonomically. Considerable quantities of chitin-decomposers live in lakes, where they probably help to mineralize the debris of zooplankton /51/.

d) An important new area of research was opened with the discovery around 1940 that CLB also live in the sea /63, 64/. This field is still only rudimentarily developed. The first marine CLB to be discovered were agar-decomposers, but there are many more types present, and those in enormous populations /28, 36, 40, 57/. The new organisms were described as Cytophaga species, so that with them the original concept of the genus was extended for the first time to non-cellulolytic organisms. At this point it became evident that in nature there are respectable populations of CLB other than the cellulose-decomposers. Different aspects of the agar-degrading CLB were investigated in several laboratories /32, 42, 63, 64, 67/. A number of species have been defined, almost all of them marine, but one, Cy. flevensis, from freshwater/66/. Since the latter organism has been found in the IJsselmeer in Holland, a former North Sea bay which gradually became freshwater after its separation from the open sea in 1932, it could well be a marine bacterium that gradually became adapted to freshwater. Support for this hypothesis comes from two observations: a) agarolytic CLB are not normally found in soil and freshwater; and b) Cy. flevensis does not contain the flexirubin-type pigments which are typical for freshwater CLB, but are usually absent from marine organisms.

e) An entirely different set of CLB was discovered in 1945 with the organisms classified in the new genus Flexibacter /60, 61/. Later it turned out that not all bacteria collected in this genus are really closely related /52/. There remain, however, very characteristic and easily recognizable organisms that are common in soil and freshwater, yet do not appear at all in the latest edition of Bergey's Manual of Determinative Bacteriology. A second genus with similar, but marine bacteria, Microscilla, was created a few years later in 1951 /50/. Contrary to an earlier stated opinion /53/, we no longer feel that the freshwater and the marine organisms are related closely enough to be united in one genus.

f) Whereas all CLB discovered thus far were strict aerobes, by 1955 the first facultatively anaerobic CLB was isolated /3/. Somewhat later, several more examples became known /2, 67/, but still no strictly anaerobic free-living CLB have been found.

g) There are, however, anaerobic CLB living in the human oral cavity and probably elsewhere. It seems that organisms of this type had been already observed at the beginning of the century, but were mistaken for fusiform bacteria. Their creeping motility has been known since at least 1940 /49/, but their relationship to the CLB was not recognized before 1961, when a new genus, Sphaerocytophaga, was created for them /23/. These are perhaps the same organisms that have recently become popular under the name Capnocytophaga /39/. The latter organisms were shown to depend on an elevated $CO_2$-level rather than on truely anaerobic conditions. While these bacteria normally seem to be harmless, they may occasionally contribute to certain disorders of the teeth, like periodontosis.

h) Around 1960 it became apparent that large populations of saprophytic CLB live in freshwater, particularly in lakes /24, 25/. Growth of these organisms usually becomes inhibited at temperatures above $25^0$. There is no question that these bacteria play an important role in the degradation of polymeric organic material in freshwater bodies /26, 51, 55/, including sewage plants /4, 27, 31/.

i) In 1963 the first CLB with a high GC-content (63 %) were reported /16/. In the following years many more CLB with such high GC-values became known, several of them with interesting properties for the applied microbiologist: They were producers of new antibiotics /18, 48/ and of unusual proteolytic enzymes /17, 22/, or lysed blue-green bacteria in water blooms /58/. The investigators dealing with them were at a loss as to the taxonomic position of the new bacteria and called them myxobacter and Sorangium, both decidedly misnomers. In the end a new genus, Lysobacter, emerged for these organisms /12/. Although close in their GC-range to that of the myxobacteria, these two groups of organisms seem to have little or nothing in common and are easily distinguished even in the vegetative state.

j) From 1960 on it became evident to several investigators that there might be connections between the CLB and some of the flavobacteria with a low GC-content /28, 41, 42, 56, 57/. A first result of this notion was a symposium on myxobacteria and flavobacteria in 1968. The idea has since found more experimental support, particularly in the field of chemosystematics /9, 20, 54/, but it is still open, what exactly these connections are, and what taxonomic measures should be taken. After all, the main purpose of this meeting is to review the progress and to stimulate further research.

k) Impressive as the long list of CLB presented above may be, there is little doubt, that upon careful examination of the diverse natural habitats still more organisms of this kind will be found /30, 40/.

## Taxonomy

Initially most investigators were impressed by the similarities between the CLB and the myxobacteria, and for a long time the two groups were freely intermingled. Thus Sporocytophaga was once included in the family Myxococcaceae, because it forms microcysts, and in the Russian scientific literature CLB are often called Promyxobacterium. Although there are data that seem to indicate some kind of relationship between the CLB and the myxobacteria (presence of menaquinones, fatty acid patterns, morphogenetic potential), other observations, as the GC-data and the unique process of fruiting body formation in myxobacteria, probably rule out a close connection. Current knowledge would justify keeping the CLB in an order of their own, Cytophagales, with the definition given above.

Further classification becomes more difficult, however. Sporocytophaga, forming microcysts; Lysobacter, with its high GC-content /12/; and Sphaerocytophaga/Capnocytophaga, anaerobic specialists living in warm-blooded animals /23, 39/, may be distinguished with sufficient accuracy and may be taxonomically meaningful groups.

The real problems begin with the genera Cytophaga and Flexibacter, in which we find at present almost all organisms listed above /11, 38/. Already a brief consideration of the metabolic and ecological diversity of all these bacteria casts serious doubts on whether this classification is reasonable. Suspicion is nourished by the distribution of the GC-data of representative strains, and becomes a certainty through DNA reconstitution experiments: In this assembly of organisms are hidden many genera and species /6, 30/. In our opinion this formidable classification problem can only be solved by first identifying relationships between strains with the methods of molecular taxonomy, and then by finding distinguishing phenomenological characters for the established taxa. It hardly seems meaningful at the moment to distinguish between different species.

One point, however, could already now be definitely improved on the basis of available data: the distinction between Cytophaga and Flexibacter, as it is proposed in the latest edition of Bergey's Manual /38/. The distinguishing character, the ability to degrade polysaccharides, seems not to separate the bacteria according to their natural relationships; there are, e.g., clear discrepancies in the GC-data /6, 30/. However, there live in soil, decaying plant material, and freshwater large populations of bacteria that correspond exactly the original description of certain Flexibacter species /53, 59, 60, 61/. These organisms have a characteristic cell shape and motility and go, in addition, through a cyclic shape from long agile thread cells in young cultures to short immotile rods in old ones. In the thread cell stage they can reliably be recognized under the microscope. The bacteria, that have been totally omitted from the latest edition of Bergey's Manual, also differ from ordinary CLB in their distinct GC-range around 50 % /6, 59/, a special modification of flexirubin-type pigments /54/, and several physiological properties /30/. We therefore suggest going back to the original definition of the genus Flexibacter, and using the said strains as neotypes.

The remaining organisms are not readily classifiable at the moment. As the hybridization and GC-data indicate that there are several or many different genera, it might be best to return to the original definition for the genus Cytophaga as well, and to reserve it for aerobic gliding cellulose-decomposers in soil and freshwater that do not form microcysts. For all other organisms new genera should be created after their relations have been sufficiently investigated.

## References


/1/ Anderson, J.I.W., Conroy, D.A.: J. Appl. Bacteriol. 32, 30-39 (1969)

/2/ Anderson, R.L., Ordal, E.J.: J. Bacteriol. 81, 130-138 (1961)

/3/ Bachmann, B.J.: J. Gen. Microbiol. 13, 541-551 (1955)

/4/ Bauer, L.: Arch. Hyg. Bakteriol. 146, 392-400 (1962)

/5/ Becker, C.D., Fujihara, M.P.: The bacterial pathogen Flexibacter columnaris and its epizootiology among Columbia River fish., Washington D.C.: Am. Fish. Soc. Monograph No. 2 (1978), 92 pp.

/6/ Behrens, H.: Charakterisierung der DNA gleitender Bakterien der Ordnung Cytophagales. Doktor Diss., TU Braunschweig, FRG (1978), 107 pp.

/7/ Borg, A.F.: Wildlife Dis. 8, 1-85 (1960)

/8/ Bredt, W., Radestock, U.: J. Bacteriol. 130, 937-938 (1977)

/9/ Callies, E., Mannheim, W.: Internat. J. Syst. Bacteriol. 28, 14-19 (1978)

/10/ Charpentier, M.: Ann. Inst. Pasteur 115, 497-537 (1968)

/11/ Christensen, P.: Can. J. Microbiol. 23, 1599-1653 (1977)

/12/ Christensen, P., Cook, F.D.: Internat. J. Syst. Bacteriol. 28, 367-393 (1978)

/13/ Colgrove, D.J., Wood, J.W.: Occurrence and control of Chondrococcus columnaris as related to Fraser River sockeye salmon. New Westminster, Brit. Columb.: Internat. Pacific Salmon Fish. Commiss. Progr. Report No. 15 (1966), 51 pp.

/14/ Collins, V.G.: Water Treatm. Examin. 19, 3-31 (1970)

/15/ Davis, H.: U.S. Bur. Fish. Bull. 38, 261-280 (1922)

/16/ De Ley, J., van Muylem, J.: Antonie van Leeuwenhoek 29, 344-358 (1963)

/17/ Ensign, J.C., Wolfe, R.S.: J. Bacteriol. 91, 524-534 (1966)

/18/ Evans, J.R., Napier, E.J., Fletton, R.A.: J. Antibiot. 31, 952-958 (1978)

/19/ Fåhraeus, G.: Symb. Bot. Upsal. 9, 1-128 (1947)

/20/ Fautz, E., Rosenfelder, G., Grotjahn, L.: J. Bacteriol. 140, 852-858 (1979)

/21/ Garnjobst, L.: J. Bacteriol. 49, 113-128 (1945)

/22/ Gillespie, D.C., Cook, F.D.: Can. J. Microbiol. 11, 109-118 (1965)

/23/ Gräf, W.: Arch. Hyg. Bakteriol. 145, 405-459 (1961)

/24/ Gräf, W.: Arch. Hyg. Bakteriol. 146, 114-125 (1962)

/25/ Gräf, W., Stürzenhofecker, P.: Arch. Hyg. Bakteriol. 148, 79-96 (1964)

/26/ Güde, H.: Arch. Hydrobiol./Suppl. 42, 483-496 (1973)

/27/ Güde, H.: Appl. Environment. Microbiol. 39, 756-763 (1980)

/28/ Hayes, P.R.: J. Gen. Microbiol. 30, 1-19 (1963)

/29/ Henrichsen, J.: Bacteriol. Rev. 36, 478-503 (1972)

/30/ Hirsch, I.: Beiträge zur Taxonomie der Cytophagales auf Grund morphologischer und stoffwechselphysiologischer Eigenschaften. Dissertation, TU Braunschweig, FRG (1980), 142 pp.

/31/ Hofsten, B. von, Berg, B., Beskow, S.: Arch. Mikrobiol. 79, 69-79 (1971)

/32/ Humm, H.J.: Duke Univ. Mar. Stat. Bull. 3, 45-75 (1946)

/33/ Hutchinson, H.B., Clayton, J.: Agric. Sci. 9, 143-173 (1919)

/34/ Imshenetsky, A.A.: Mikrobiologie der Cellulose. Berlin: Akademie Verlag (1959); 466 pp.

/35/ Johnson, D.: J. Bacteriol. 24, 335-340 (1932)

/36/ Kadota, H.: Memoirs College Agric., Kyoto Univ. No. 74, 1-128 (1956)

/37/ Krzemieniewska, H.: Arch. Mikrobiol. 4, 394-408 (1933)

/38/ Leadbetter, E.R.: Cytophagales, in Buchanan, R.E., Gibbons, N.E., Edts., Bergey's manual of determinative bacteriology, 8th ed., Baltimore: The Williams and Wilkins Comp. (1974), pp. 99-127

/39/ Leadbetter, E.R., Holt, S.C., Socransky, S.S.: Arch. Microbiol. 122, 9-16 (1979)

/40/ Lewin, R.A.: J. Gen. Microbiol. 58, 189-206 (1969)

/41/ McMeekin, T.A., Patterson, J.T., Murray, J.G.: J. Appl. Bacteriol. 34, 699-716 (1971)

/42/ Mitchell, T.G., Hendrie, M., Shewan, J.M.: J. Appl. Bacteriol. 32, 40-50 (1969)

/43/ Norman, A.G., Fuller, W.H.: Edv. Enzymol. 2, 239-264 (1942)

/44/ Ordal, E.J., Rucker, R.R.: Proc. Soc. Exp. Biol. Med. 65, 15-18 (1944)

/45/ Ostertag, H.: Ergebn. Hyg. Bakteriol. Immunitätsforsch. Exp. Ther. 27, 149-322 (1952)

/46/ Pate, J.L., Chang, L.Y.E.: Curr. Microbiol. 2, 59-64 (1979)

/47/ Pate, J.L., Ordal, E.J.: J. Cell Biol. 35, 37-51 (1967)

/48/ Peterson, E.A., Gillespie, D.C., Cook, F.D.: Can. J. Microbiol. 12, 221-230 (1966)

/49/ Prêvot, A.R.: C.R. Soc. Biol. 133, 246-249 (1940)

/50/ Pringsheim, E.G.: J. Gen. Microbiol. 5, 124-149 (1951)

/51/ Reichardt, W.: Verh. Internat. Verein. Limnol. 19, 2636-2642 (1975)

/52/ Reichenbach, H., Golecki, J.R.: Arch. Microbiol. 102, 281-291 (1975)

/53/ Reichenbach, H., Kleinig, A., Achenbach, H.: Arch. Microbiol. 101, 131-144 (1974)

/54/ Reichenbach, H., Kohl, W., Böttger-Vetter, A., Achenbach, H.: Arch. Microbiol. 126, 291-293 (1980)

/55/ Ruschke, R., Rath, M.: Arch. Hydrobiol./Suppl. 28, 377-402 (1966)

/56/ Shewan, J.M.: J. Appl. Bacteriol. 32, 1-4 (1969)

/57/ Shewan, J.M.: A. Appl. Bacteriol. 34, 299-315 (1971)

/58/ Shilo, M.: J. Bacteriol. 104, 453-461 (1970)

/59/ Simon, G.D., White, D.: Arch. Mikrobiol. 78, 1-16 (1971)

/60/ Soriano, S.: Rev. Argent. Agron., Buenos Aires 12, 120-140 (1945)

/61/ Soriano, S.: Antonie van Leeuwenhoek 12, 215-222 (1947)

/62/ Soriano, S., Lewin, R.A.: Antonie van Leeuwenhoek 31, 66-80 (1965)

/63/ Stanier, R.Y.: J. Bacteriol. 40, 619-636 (1940)

/64/ Stanier, R.Y.: J. Bacteriol. 42, 527-558 (1941)

/65/ Stanier, R.Y.: J. Bacteriol. 53, 297-315 (1947)

/66/ Van der Meulen, H.J., Harder, W., Veldkamp, H.: Antonie van Leeuwenhoek 40, 329-346 (1974)

/67/ Veldkamp, H.: J. Gen. Microbiol. 26, 331-342 (1961)

/68/ Winogradsky, S.: Ann. Inst. Pasteur 43, 549-633 (1929)

# EMENDATION OF THE GENUS FLAVOBACTERIUM AND THE STATUS OF THE GENUS. DEVELOPMENTS AFTER THE 8TH EDITION OF BERGEY'S MANUAL

B. Holmes, R.J. Owen

National Collection of Type Cultures, Central Public Health Laboratory, Colindale Avenue, London NW9 5HT, UK

Summary

In the 8th edition of Bergey's Manual [25] the genus Flavobacterium remained taxonomically heterogeneous, being divided into two sections. Section I contained six non-motile species with the G+C content of their DNA in the range 30 to 42 mol % (low G+C content strains) and Section II contained six species that were either non-motile or motile and peritrichous but with G+C contents in the range 63 to 70 mol % (high G+C content strains). Here we describe the general background and sequence of events, culminating in an emended description of the genus [12], in which restriction of Flavobacterium to certain well-described low G+C content species is proposed. The reasons for replacing the current type species F. aquatile, with F. breve will also be discussed. Although only four species conform to the emended description of Flavobacterium at present, other taxa that are candidates for inclusion in the genus will be described as well as the taxonomic problems associated with them; particular reference will be made to Flavobacterium Group IIb [23]. The taxonomic status of the high G+C content strains of Section II will be discussed as will the status of certain taxa, such as Group IIf [23], which are sometimes referred to as Flavobacterium-like.

## Introduction

The genus Flavobacterium in the 8th edition of Bergey's Manual [25] was taxonomically heterogeneous, being divided into two sections. Section I contained the six non-motile species shown in Table 1 a) and all these had low G+C contents in the range 30 - 42 mol % G+C. Of the species included in Section I, only Flavobacterium meningosepticum was described as being of clinical interest. This organism has been well documented as a cause of meningitis in the newborn [16]. F. meningosepticum was adequately described and there are several reference cultures of the species available, unlike most other *Flavobacterium* species which are represented by only a single strain. *F. meningosepticum* is seen only occasionally in the specimens received in our laboratory; but the organism designated *Flavobacterium* Group IIb by Tatum and co-workers [23] is more often received. Although these two taxa are difficult to separate on routine biochemical tests they can be differentiated according to the degree of pigment that they produce; strains of Group IIb produce a deep yellow pigment whilst strains of *F. meningosepticum* produce only a pale yellow pigment and may often be non-pigmented. From the strains received in our laboratory for identification, the concept of *Flavobacterium* developed as comprising Gram-negative, non-motile, non-gliding, non-spreading strains with low G+C contents. The reservation of *Flavobacterium* for those organisms in Section I of Bergey 8 [25] thus seemed appropriate.

TABLE 1

The *Flavobacterium* species included in Bergey 8 [25].

| | | | |
|---|---|---|---|
| a) | Section I: | | |
| | *F. aquatile* | *F. meningosepticum* | *F. halmephilum* |
| | *F. breve* | *F. ferrugineum* | *F. uliginosum* |
| b) | Section II: | | |
| | *F. capsulatum* | *F. rigense* | *F. tirrenicum* |
| | *F. lutescens* | *F. indoltheticum* | *F. devorans* |

The six species listed in Table 1 b), are included in *Flavobacterium* Section II of Bergey 8. These species have a high G+C content, and are either non-motile or motile and peritrichous, but none were reported to occur in clinical material [25]. It thus seemed inappropriate to suggest that *Flavobacterium* be reserved solely for the high G+C content species with the low G+C content species classified in other genera such as *Cytophaga* or *Empedobacter*. Since the majority of microbiologists are employed in the clinical field, least confusion would result if the name *Flavobacterium* were reserved for that group of organisms which contains the clinically important species such as *F. meningosepticum*. Later studies, however, were to show that two of the species contained in Section II in fact were low G+C content species which should have been included in Section I and it also appeared that two of the other species in Section II apparently did occur in clinical material (see below).

## Revised descriptions of F. odoratum and F. breve

One of the major problems with *Flavobacterium* is that many of the species ascribed to the genus are represented by only a single strain making it difficult to determine their taxonomic position. Recently, for example, the new species *Flavobacterium oceanosedimentum* [6] was proposed on just a single strain. If the taxonomic confusion surrounding *Flavobacterium* is ever to be clarified then species each based on the description of several reference strains are necessary.

Some reference strains of named species of Flavobacterium were obtained in order to compare with our hitherto unclassified Flavobacterium-like isolates. An initial discovery was that some of the clinical isolates received for identification in our laboratory corresponded to the ATCC reference strain (ATCC 4651) of Flavobacterium odoratum. Altogether nine field isolates were identified to the species and on the basis of these ten strains a revised description of the species was provided [14] and three reference strains were deposited in the NCTC. The ATCC reference strain, which originated from Stutzer, the original describer of F. odoratum, was designated the neotype strain. This strain became the type strain on publication of the Approved Lists of Bacterial Names [22]. F. odoratum was also found to correspond to another of the unnamed groups recognised by Tatum and co-workers [23] and in this case it was Group M-4f.

Again, study of the hitherto unclassified Flavobacterium isolates received by the NCTC revealed that several corresponded to the ATCC reference strain (ATCC 14234) of F. breve. A revised description of this species based on seven strains was provided [15] and three reference strains were deposited in the NCTC. The ATCC reference strain of F. breve, which came from a fish, was not chosen as the neotype strain because it was slightly atypical of the species. A more typical strain, NCTC 11099, was chosen as the neotype and now there are four strains of this species available from culture collections.

## A homogeneous group within Flavobacterium

From Table 2 it can be seen that among the various taxa that now conform to Week's Section I a group of five taxa found in clinical material was discernible. This comprised three named species (F. breve, F. meningosepticum and F. odoratum), with Group IIb mentioned above and also Group IIk, type 2. In an earlier study of Group IIk [13] it had been established that Group IIk, type 1 belonged in the genus Pseudomonas as it is currently defined. The Group IIk, type 2 strains examined had lower DNA base compositions (39.6 mol % G+C), which were in the range of values of those species classified in Flavobacterium Section I. Strains of each of the five taxa from clinical material were yellow-pigmented and resistant to therapeutic levels of many antimicrobial agents. This resistance to antimicrobial agents appears to be controlled by chromosomal genes in these organisms and is not a plasmid mediated characteristic [18]. From the work of Callies and Mannheim [4], four of the five taxa were known to contain menaquinones whereas those taxa not known to occur in clinical material were more heterogeneous with two of the species containing ubiquinones and two menaquinones. Because the type species of Flavobacterium, F. aquatile, contains ubiquinones, Callies and Mannheim [4] considered that Flavobacterium should be restricted to those organisms resembling F. aquatile and containing ubiquinones. We would not share such a view as it could result in a change of generic name for those species of clinical interest currently classified in Flavobacterium, thereby causing confusion to many microbiologists.

TABLE 2

Low G+C content Flavobacterium species

| Taxa occuring in clinical material | | Taxa not occuring in clinical material | |
|---|---|---|---|
| F. breve (MK)* | Group IIb (MK) | F. aquatile (Q) | F. halmephilum (Q) |
| F. meningosepticum (MK) | Group IIk, type 2 (NT) | F. ferrugineum (MK) | F. uliginosum (MK) |
| F. odoratum (MK) | | | |

* MK = menaquinones, Q = ubiquinones, NT = not tested; data from Callies and Mannheim [4].

## Replacement of F. aquatile as type species

For many reasons it seemed unlikely that F. aquatile could belong in the same genus as those Flavobacterium species of clinical interest. In order not to confuse medical microbiologists by a possible change of genus for the clinically important Flavobacterium species it seemed prudent to consider the possibility of replacing F. aquatile as the type species of the genus. F. aquatile is a species represented just by a single strain and this particular strain has a cell wall structure similar to that of Cytophaga johnsonae [9]. The strain was also able to swarm after prolonged incubation at 15°C [17] and Perry claimed to demonstrate gliding cellular motility [21]. There is thus some evidence that F. aquatile is an atypical Cytophaga and if this is true, then because Flavobacterium was named prior to Cytophaga it could mean that all those species currently classified in Cytophaga might end up classified in Flavobacterium. There is also evidence that the reference strain of F. aquatile is neither a typical Flavobacterium nor a typical Cytophaga. The strain remained ungrouped in a numerical phenetic analysis of Cytophaga and Flavobacterium strains carried out by Hayes [11] and we have already mentioned that the strain contains ubiquinones [4] unlike the typical Cytophaga and low G+C content Flavobacterium species.

We decided to prepare a Request for an Opinion [12] to effect the removal of F. aquatile as the type species of the genus and F.breve seemed the logical choice for a replacement type species. F. breve was included in the original description of Flavobacterium [3]. Apart from F. aquatile it was the only original species retained in Bergey 8 [25]. The species description had recently been revised [15] including a proposal of a neotype strain. The species was more typical of a group of similar bacteria. There are certainly more strains of the species available for study and Perry (personal communication) has been unable to demonstrate evidence of swarming or gliding motility which might suggest an affinity to the cytophagas. Flavobacterium breve was not included on the Approved Lists of Bacterial Names [22] and we shall ensure the name is revived.

## Taxa conforming to the emended description of Flavobacterium

The description of the genus Flavobacterium was emended [12] to make the genus more taxonomically homogeneous. Only four named species corresponded to the emended description (Table 3). As well as F. breve, F. meningosepticum and F. odoratum as previously mentioned., F. balustinum [10] was also found to fit the description. However, apart from these named species, there are also other taxa which have to be

TABLE 3

Taxa conforming to the emended description of Flavobacterium

| | | |
|---|---|---|
| F. balustinum | Group IIb | Group IIf |
| F. breve | Group IIk, type 2 | Group IIj |
| F. meningosepticum | "Group IIk, type 3" | |
| F. odoratum | | |

considered and these include Group IIb, as previously mentioned, and Group IIk, type 2. After the Request for an Opinion [12] it seemed logical to consider nomenclatural proposals for Groups IIb and IIk, type 2. The name Flavobacterium multivorum for Group IIk, type 2 has thus been proposed [Holmes, Owen and Weaver, unpublished results]. A numerical taxonomic analysis of Group IIk, type 2 strains [Holmes and Owen, unpublished results] showed a new group comprising only two strains but a third isolate has since been found and it seems likely that this will form yet another new species referred to in Table 3 as "CDC Group IIk, type 3". As soon as some additional isolates have been found, a name will be proposed for this species. Nomenclatural proposals for Group IIb are under consideration but here difficulty is encountered because the group is heterogeneous in its DNA characteristics [20] and may contain two or three different species. The description of F. balustinum needs to be revised and any such description should be given in conjunction with that for Group IIb. The two taxa are biochemically indistinguishable although F. balustinum has a lower DNA base composition than the range of typical Group IIb strains. CDC Groups IIf and IIj conform to the emended description of the genus but there are reasons for considering that they should be placed in a genus separate from Flavobacterium (see below).

## Future developments in the taxonomy of Flavobacterium

It seems that the organisms conforming to the emended description of Flavobacterium [12] might eventually constitute four prospective genera (Table 4). Most of the species of clinical interest would be retained in prospective genus A i.e. Flavobacterium breve, F. meningosepticum and Group IIb. Prospective genus B would include only F. odoratum, and this differs from genus A: it is non-saccharolytic and fails to produce indole. Genus C contains CDC Group IIk, type 2 as well as the organism herein referred to as "Group IIk, type 3". These organisms are saccharolytic like those in genus A but they fail to produce indole and are also non-proteolytic. The three genera A, B and C are resistant to antimicrobial agents whilst genus D is much more susceptible. It is genus D that would include CDC Groups IIf and IIj. Such organisms probabably should not be included in Flavobacterium even though they do produce indole and are proteolytic. Their greater susceptibility to antimicrobial agents, their non-saccharolytic nature, their failure to produce yellow pigment, the smaller genome size confirmed for Group IIf, and the fact that they tend to be strict parasites rather than free-living, may possibly exclude them from the genus Flavobacterium as currently defined [12]. Owen and Snell [19] compared Group IIf with Flavobacterium and Moraxella and failed to reach any definite conclusion as to which of these genera Group IIf might be ascribed. In DNA base composition, ability to produce indole, ability to grow well on nutrient agar and ability to produce a brown melanin-like pigment on tyrosine agar, Group IIf shows close affinity to Flavobacterium. However, the non-saccharolytic nature of Group IIf and its susceptibility to penicillin suggest affinity to Moraxella.

TABLE 4

Differentiation of the prospective genera*

| Character | Prospective genus | | | |
|---|---|---|---|---|
| | A | B | C | D |
| Yellow pigment | +/- | + | + | - |
| Saccharolytic activity | + | - | + | - |
| Indole production | + | - | - | + |
| Proteolytic activity | + | + | - | + |
| Antimicrobial susceptibility | R | R | R | S |

* A contains F. balustinum, F. breve, F. indoltheticum, F. meningosepticum and Group IIb; B contains F. odoratum; C contains Group IIk, type 2 and "Group IIk, type 3"; D contains Group IIf and Group IIj.

Thus each of the four broad groups amongst the microorganisms, which conform to the emended definition of Flavobacterium [12], may warrant recognition as separate genera at some future date. There is no real evidence to support this view but should it prove to be correct then it is important that Flavobacterium should be reserved for genus A. Even though organisms such as Group IIk, type 2 may eventually find a home in a separate genus, it is preferable at present to include them in Flavobacterium at least until there is good evidence to the contrary. It is certainly preferable to name such organisms so that they become more widely known and studied rather than to continue referring to them simply by the designations given by Tatum and co-workers [23]. The problem with proposing names for such groups, however, is that they may be heterogeneous and be comprised of more than one species. F. indoltheticum, which was included in Section II of Bergey 8 [25], is in fact a low G+C content strain, and its taxonomic position will only be resolved in conjunction with studies of F. balustinum and Group IIb. There is, however, some evidence to suggest that F. indoltheticum may be more closely aligned to F. breve. It is known from previous DNA studies that Group IIb, which is phenotypically very heterogeneous, can be divided into two or three separate DNA relatedness groups [20]. It was hoped therefore that our numerical taxonomic studies would yield phenotypic data to correlate with the DNA relatedness groups so that these might be recognised as separate species. A recent numerical taxonomic analysis [Holmes and Owen, unpublished results] shows that Group IIk, type 2 and F. odoratum are distinct taxa. F. meningosepticum was distinct from Group IIb, but within the two taxa there was heterogeneity with the formation of several sub-clusters; the type strain of F. balustinum was contained in the Group IIb cluster. In this analysis, F. breve was barely distinguishable from Group IIb, and the type strain of F. indoltheticum was contained in the F. breve cluster. Further analysis is necessary before we can draw any definite conclusions but, it appears that there is little phenotypic data to correlate with the DNA relatedness groups in Group IIb. The latter taxon may have to be recognised as a single species for which the name F. balustinum may be appropriate.

## The high G+C content "Flavobacterium" strains

Table 1 b) lists those taxa included in Section II of Bergey 8 [25]; of these probably F. capsulatum, and certainly F. devorans, could be accommodated in Pseudomonas as that genus is currently defined. F. devorans is a synonym of P. paucimobilis [12] and the same may also be true of F. capsulatum. This problem of synonyms warrants further study as there are two or three Pseudomonas species which also appear to be synonyms of P. paucimobilis. F. indoltheticum belongs in Section I. F. lutescens belongs in the same rRNA superfamily [2] as species of Pseudomonas Section I of Bergey 8 [8]. F. tirrenicum apparently should also be included in Section I of Flavobacterium [2]. Thus, the concept of Flavobacterium Section II was virtually defunct prior to publication of the Approved Lists [22]. With the exception of F. rigense, it appeared that all the other five species included in Section II could have been accommodated elsewhere so that suggestion of reviving the generic name Empedobacter for Section II would have been unnecessary. However, the classification of F. lutescens and F. tirrenicum is now academic as these two species, together with F. rigense, do not appear on the Approved Lists [22] so have no further standing in nomenclature unless revived.

## The Approved Lists as applied to Flavobacterium

The Approved Lists [22] as they relate to Flavobacterium proved to be a disappointment. It seemed that the opportunity to clear out species of uncertain classification was not taken. For the genus Flavobacterium to be homogeneous a number of species now on the Approved Lists will have to be transferred elsewhere. Table 5 lists the species which appear on the Approved Lists and can be regarded as belonging to Section I. F. balustinum, F. meningosepticum and F. odoratum are three of the named species for which it has been proposed the generic name Flavobacterium should be conserved [12]. Two of the other species retained on the Approved Lists are ones that were considered species incertae sedis in Bergey 8 [25]. The species listed in Table 5, excluding the species of clinical interest, are heterogeneous. F. acidificum, although excluded from Flavobacterium [12] as Gram-positive [25], appears to be Gram-negative and the type strain is considered to be a strain of Erwinia herbicola subsp. ananas [1]. Rejection of F. aquatile as a nomen dubium has been proposed [12]. Three species have similar G+C contents; F. ferrugineum (48.6 moles % G+C [12]), F. halmephilum (49.7 moles % G+C [4]) and F. heparinum (45.6 moles % G+C [21]). They could form a separate group, except F. halmephilum belongs [2] in the same rRNA

TABLE 5

Low G+C content Flavobacterium species appearing on the Approved Lists of Bacterial Names [22]

| | | |
|---|---|---|
| F. balustinum (NT)* | F. acidificum† (Q) | F. ferrugineum (MK) |
| F. meningosepticum (MK) | F. aquatile (Q) | F. halmephilum (Q) |
| F. odoratum (MK) | F. uliginosum (MK) | F. heparinum † (MK) |

*MK = menaquinones, Q = ubiquinones, NT = not tested; data from Callies and Mannheim [4]. † Species incertae sedis in Bergey 8 [25].

superfamily as species of Pseudomonas Section I in Bergey 8 [8]; Callies and Mannheim [4] have shown this species contains ubiquinones. In contrast F. ferrugineum and F. heparinum contain menaquinones and F. heparinum has recently been transferred to Cytophaga as Cytophaga heparina [7]. F. uliginosum is the only other species in Table 5 which has a DNA base composition in the range 30 - 42%. This species has been excluded from the genus Flavobacterium because of its agarolytic activity [12]. It is possible that F. uliginosum (32 moles % G+C [25]) might find a home with the marine agarolytic species of Cytophaga which have similar G+C contents, such as C. lytica (33.2 to 34.2 moles % G+C [7]) and C. latercula (34.2 moles % G+C [7]).

Table 6 lists those Flavobacterium species on the Approved Lists [22] which might be regarded as belonging to Section II. F. capsulatum and F. devorans have already been mentioned as possible synonyms of Pseudomonas paucimobilis and F. indoltheticum may belong to Group IIb or to the F. breve group. Three of the species were considered to be species incertae sedis in Bergey 8 [25].

TABLE 6

High G+C content "Flavobacterium" species appearing on the Approved Lists of Bacterial Names [22]

| | | | |
|---|---|---|---|
| F. capsulatum | F. esteroaromaticum* | F. acidurans | F. okeanokoites |
| F. devorans | F. marinotypicum* | F. oceanosedimentum | |
| F. indoltheticum | | F. resinovorum* | |

* Species incertae sedis in Bergey 8 [25].

F. esteroaromaticum and F. marinotypicum are Gram-positive [25] and F. esteroaromaticum is probably a coryneform [2]. Three of the remaining species have high G+C contents: F. acidurans (66.3 % G+C [12]), F. oceanosedimentum (67.5 % G+C [6]) and F. resinovorum (70 % G+C [25]), as presumably does F. okeanokoites which is described as having peritrichous flagella [12].

Those working with clinical isolates are fortunate because in the clinical environment the selective pressures are more constant and possibly result in more discrete species. However, in complex ecological situations such as soil and water the varied selection pressures may cause species to be less distinct and to form a continuous spectrum. From clinical isolates it has been possible to describe a number of species in Flavobacterium which form a relatively homogeneous group [14,15, 16]. For each species there are designated type strains, full species descriptions, and several representative strains deposited in culture collections. These should provide a frame-work within which non-clinical microbiologists may study Flavobacterium and Flavobacterium-like strains. For example, some of the environmental isolates studied by Hayes [11] correspond to F. breve whilst others correspond to Group IIb [18].

Taxonomic studies of Group IIb and "Group IIk, type 3" continue and nomenclatural proposals will be made for each. Attention will then be focussed on the Flavobacterium-like organisms such as Groups IIf and IIj. Acceptance of the proposals contained in the Request for an Opinion [12] would make the genus Flavobacterium more homogeneous than it has ever been, but eventually it may be possible to re-organize along the lines suggested above. Elucidation of the relationship of F. uliginosum with the marine agarolytic species of Cytophaga also warrants further investigation, as does the classification of the four species with high G+C contents (F. acidurans, F. oceanosedimentum, F. okeanokoites and F. resinovorum). These could possibly be included in the same genus, and the name Empedobacter might be resurrected for them. New genera would also need to be found for the Gram-positive species, as well as for species such as F. ferrugineum (48.6 moles % G+C [12]).

One major problem is the differentiation of Flavobacterium from Cytophaga. This has always been difficult, because of the different interests of workers studying Flavobacterium and studying Cytophaga. It must be hoped, that a concensus of opinion will emerge that Flavobacterium should be reserved for the low G+C content species in particular those of clinical importance. A useful initial step might be the formation of an informal working party of people who have good collections of Flavobacterium strains, and others with good collections of Cytophaga strains. Strains could then be exchanged and studied in collaborating institutions and the long standing problem of the differentiation of Cytophaga from Flavobacterium might eventually be resolved. Few workers would disagree with the recent report by Webster and Hugh [24] that Flavobacterium species are strict aerobes and therefore in the next edition of Bergey's Manual, Flavobacterium should not be considered along with fermentative organisms as it was in Bergey 8 [25]. Indeed, the recent suggestion by Callies and

Mannheim [5] that Flavobacterium should be transferred to the Cytophagaceae may be useful.

## References

[1] American Type Culture Collection: p. 77 Catalogue of Strains I, 14th ed. Rockville, Maryland (1980): American Type Culture Collection

[2] Bauwens, M. : Ant. van Leeuw. J. Microbiol. Serol. 46, 95 (1980)

[3] Bergey, D.H., Harrison, F.C., Breed, R.S., Hammer, B.W., Huntoon, F.M. (eds): p. 97-117 in: Bergey's Manual of Determinative Bacteriology. Baltimore (1923) Williams and Wilkins

[4] Callies, E., Mannheim, W. : Int. J. Syst. Bacteriol. 28, 14-19 (1978)

[5] Callies, E., Mannheim, W. : Ant. van Leeuw. J. Microbiol. Serol. 46, 41-49 (1980)

[6] Carty, C.E., Litchfield, C.D. : Int. J. Syst. Bacteriol. 28, 561-566 (1978)

[7] Christensen, P. : Int. J. Syst. Bacteriol. 30, 473-475 (1980)

[8] Doudoroff, M., Palleroni, N.J. p. 217-243 in : Buchanan, R.E., Gibbons, N.E. (eds) Bergey's Manual of Determinative Bacteriology. Baltimore (1974) Williams and Wilkins

[9] Follett, E.A.C., Webley, D.M. : Ant. van Leeuw. J. Microbiol. Serol. 31, 361-382 (1965)

[10] Harrison, F.C. : Can. J. Res. 1, 214-239 (1929)

[11] Hayes, P.R. : J. Appl. Bacteriol. 43, 345-367 (1977)

[12] Holmes, B., Owen, R.J. : Int. J. Syst. Bacteriol. 29, 416-426 (1979)

[13] Holmes, B., Owen, R.J., Evans, A., Malnick, H., Willcox, W.R. : Int. J. Syst. Bacteriol. 27, 133-146 (1977)

[14] Holmes, B., Snell, J.J.S., Lapage, S.P. : Int. J. Syst. Bacteriol. 27, 330-336 (1977)

[15] Holmes, B., Snell, J.J.S., Lapage, S.P. : Int. J. Syst. Bacteriol. 28, 201-208 (1978)

[16] King, E.O. : Am. J. Clin. Pathol. 31, 241-247 (1959)

[17] Mitchell, T.G., Hendrie, M.S., Shewan, J.M. : J. Appl. Bacteriol. 32, 40-50 (1969)

[18] Owen, R.J., Holmes, B. : this symposium

[19] Owen, R.J., Snell, J.J.S. : Ant. van Leeuw. J. Microbiol. Serol. 39, 473-480 (1973)

[20] Owen, R.J., Snell, J.J.S. : J. Gen. Microbiol. 93, 89-102 (1976)

[21] Perry, L.B. : J. Appl. Bacteriol. 36, 227-232 (1973)

[22] Skerman, V.B.D., McGowan, V., Sneath, P.H.A. (eds) : Int. J. Syst. Bacteriol. 30, 225-420 (1980)

[23] Tatum, H.W., Ewing, W.H., Weaver, R.E. : p. 270-294 in : Lennette, E.H., Spaulding, E.H., Truant, J.P. (eds), Manual of Clinical Microbiology, 2nd ed. Washington D.C. (1974) American Society for Microbiology

[24] Webster, J.A., Hugh, R.: Int. J. Syst. Bacteriol. 29, 333-338 (1979)

[25] Weeks, O.B. : p. 357-364 in : Buchanan, R.E., Gibbons, N.E. (eds), Bergey's Manual of Determinative Bacteriology. Baltimore (1974) Williams and Wilkins

# IMPROVEMENTS IN THE TAXONOMY OF *FLAVOBACTERIUM* BY DNA:rRNA HYBRIDIZATIONS

M. Bauwens and J. De Ley

Laboratorium voor Mikrobiologie en mikrobiële Genetica,
Rijksuniversiteit, Ledeganckstraat, 35, B-9000 Gent, Belgium

## Summary

DNA:rRNA hybridizations revealed the considerable heterogeneity of named flavobacteria. Low % GC *Flavobacterium aquatile*, *F. pectinovorum*, *F. tirrenicum*, *F. uliginosum*, *F. odoratum*, *F. breve*, *F. meningosepticum* and several named *Cytophaga* species form a large, heterogeneous cluster. Several other *Flavobacterium* species are misnamed and belong in entirely different taxa.

## Introduction

*Flavobacterium* was created in 1923 /1/ as a color genus. From the very beginning it was heterogeneous. In every revision it lost some of its heterogeneity. In 1957 /2/ the gram-positive species were dropped. In 1974 /15/ the non-motile species with cytophaga -like swarming were excluded. In the sixties the basic importance of the DNA base composition as a rough measure of genetic dissimilarities was convincingly demonstrated. In 1963 De Ley and Van Muylem /6/ drew attention for the first time to the great heterogeneity in the % GC values of some named *Cytophaga* and *Flavobacterium*, which extended from 33 to about 69 % GC. It was at once obvious that this group was genetically very heterogeneous. Therefore, it was also quite clear that one of the next criteria for the internal classification of *Flavobacterium* had to be the DNA base composition and that the authentic flavobacteria are low

% GC species around the type species F. aquatile or F. breve /11/.

We describe here briefly the applications of yet another molecular approach to improve the present taxonomy of this genus. We showed previously /4,7,8,10/ that the similarities between bacterial rRNA cistrons, determined by DNA:rRNA hybridizations involving numerous bacterial genera and hundreds of species, are an excellent measure to group bacteria at the generic and suprageneric levels. It is also a very useful technique to detect misnamed organisms.

## Methods

Therefore we carried out a large number of hybridizations between $^{14}C$- or tritiated rRNA of F. aquatile ATCC 11947, F. capsulatum ATCC 14666, Cytophaga johnsonae ATCC 17061, Arthrobacter oxydans CBRI 21010, Azotobacter chroococcum NCIB 8002, Azomonas agilis NCIB 8636, Pseudomonas fluorescens MMCA 40, E. coli B, "Alcaligenes" aquamarinus NCMB 557 and Alteromonas vaga ATCC 27119 and DNA of numerous species of Flavobacterium, some Cytophaga and Flexibacter as well as several other reference genera. The hybridizations were carried out according to De Ley and De Smedt /3/ and De Ley and Tytgat /5/.

## Results and Discussion

From our results it is quite evident that the named flavobacteria belong to at least six taxonomically widely different branches and that the majority of these organisms is misnamed.

1) The low % GC flavobacteria (31.6 - 39 % GC) are very heterogeņeous amongst themselves. F. aquatile, F. pectinovorum, F. tirrenicum and Cyt. johnsonae are close together as far as the three experimental criteria, $T_{m(e)}$, % rRNA binding and % GC are concerned. In the vicinity there is another small subgroup, clearly separated, consisting of Cyt. marinoflava, Cyt. lytica, Cyt. salmonicolor, F. uliginosum, Flex. aurantiacus subsp. excathedrus and perhaps McMeekin's yellow organism 42. The third subgroup consists of F. odoratum, a fourth homogeneous subgroup consists of F. breve and the fifth subgroup of F. meningosepticum. We do not want to make nomenclatural proposals here, but the authentic genus Flavobacterium will have to consist of all or some of the above named flavobacteria. It is not possible with our methods to separate the genera Cytophaga and the authentic Flavobacterium clearly from each other. The Cytophaga - Flavobacterium complex is very heterogeneous as the $\Delta T_{m(e)}$ range is 10°C, which is

as broad as the entire family of the Enterobacteraceae. Other methods such as DNA:DNA hybridizations, numerical analysis of phenotypic features and comparisons of protein gel electrophoregrams will have to be used to attempt clear and convincing differentiation. F. balustinum was not available. All the other named Flavobacterium species which we examined are misnamed and belong in entirely different taxonomic areas.

Flexibacter roseolus, Flex. ruber and Flex. aurantiacus are related amongst themselves but quite different from the above groups: they are as far removed from the Cytophaga-Flavobacterium complex as the other heterotrophic taxa in the rRNA superfamilies I to IV. Flex. aurantiacus subsp. excathedrus is an exception, as it is very close to Cyt. lytica and Cyt. marinoflava. This Flexibacter subspecies is probably misnamed.

2) Hybridizations with labeled rRNA of F. capsulatum ATCC 14666 showed that this strain is a member of our rRNA superfamily IV /8,10/, which is constituted by a number of genera such as Acetobacter, Gluconobacter, Agrobacterium, Rhizobium, Spirillum, Azospirillum, etc. The closest relatives of "F." capsulatum are a few yellow pigmented organisms such as "Ps." azotocoligans, "Chr." folium, "Ps." paucimobilis and "F." devorans. These four taxa lay at about -6° $\Delta T_{m(e)}$ from "F." capsulatum. It is very likely that they are closely related. Yabuuchi et al. /16/ recently proposed that F. devorans is identical to Ps. paucimobilis, which is supported by the present data.

3) Hybridizations with labeled rRNA of Arthrobacter oxydans CBRI 21010. From previous experience on DNA:rRNA hybrids (De Smedt and De Ley, to be published) it appeared that the entire group of Arthrobacter, Corynebacterium, Brevibacterium, Microbacterium, Nocardia, etc. constitutes a separate rRNA superfamily which is very heterogeneous. We found that the following misnamed flavobacteria belong in this superfamily: "F." dehydrogenans, "F." arborescens, "F." esteroaromaticum, "F." flavescens and "F." suaveolens. The strain "F." aquatile CCM 1948 is doubly misnamed since it is neither a Flavobacterium nor an aquatile but it belongs here. A representative of pleiston A of Floodgate and Hayes /9/ and the yellow organisms no 13 and 48 of McMeekin /12/ also belong here. All these organisms are rather far removed from Arthrobacter oxydans with a $\Delta T_{m(e)}$ of 9 to 15° C. Broadly speaking it seems that they belong in the coryneforms or are related with them. The detailed taxonomic position of all these yellow organisms remains to be established.

4) The taxonomic position of "F." lutescens and "F." denitrificans From hybridizations with labeled rRNA from Azomonas, Azotobacter and

Pseudomonas fluorescens it is evident that both misnamed flavobacteria belong in the genus Pseudomonas, in particular in the large Pseudomonas section I. Their exact specific position remains to be established. Our results disagree with the proposal /3/ to rename the former organism as Empedobacter lutescens.

5) The taxonomic position of "F." (Obesumbacterium) proteus. Priest et al. /14/ included these organisms as Hafnia in the Enterobacteraceae. Our hybridizations with labeled rRNA from E. coli confirmed the position of "F." proteus as a member of this family.

6) The taxonomic position of "F." halmephilum. Recently we discovered in this laboratory the existence of a separate group of marine, peritrichously flagellated, gram-negative rods with high % GC. We shall elsewhere give separate genus status to this group. Hybridizations with rRNA of the reference "Alcaligenes" aquamarinus strain showed that "F." halmephilum is either a member of this group of organisms or else a very close relative.

7) We established that "F." heparinum and "F." ferrugineum are possibly border cases and "F." okeanokoites does not belong in the Flavobacterium-Cytophaga complex. They are thus misnamed. We do not know their exact taxonomic position.

## References


/1/ Bergey, D.H., Harrison, F.C., Breed, R.S., Hammer, B.W., Huntoon, F.M. (ed.), Bergey's Manual of Determinative Bacteriology, 1st ed. Baltimore (1923): The Williams and Wilkins Co.

/2/ Breed, R.S., Murray, E.G.D., Smith, N.R. (ed.), Bergey's Manual of Determinative Bacteriology, 7th ed. Baltimore (1957): The Williams and Wilkins Co.

/3/ De Ley, J., De Smedt, J.: Ant. van Leeuwenh. J. Microb. 41, 287-307 (1975)

/4/ De Ley, J., Segers, P., Gillis, M.: Int. J. Syst. Bacteriol. 28, 154-168 (1978)

/5/ De Ley, J., Tytgat, R.: Ant. van Leeuwenh. J. Microb. 36, 461-474 (1970)

/6/ De Ley, J., Van Muylem, J.: Ant. van Leeuwenh. J. Microb., 29, 344-358 (1963)

/7/ De Smedt, J., Bauwens, M., Tytgat, R., De Ley, J.: Int. J. Syst. Bacteriol. 30, 106-122 (1980)

/8/ De Smedt, J., De Ley, J.: Int. J. Syst. Bacteriol. 27, 222-240 (1977)

/9/ Floodgate, G.D., Hayes, P.R.: J. Gen. Microb. 30, 237-244 (1963)

/10/ Gillis, M., De Ley, J.: Int. J. Syst. Bacteriol. 30, 7-27 (1980)

/11/ Holmes, B., Owen, R.J.: Int. J. Syst. Bacteriol. 29, 416-426 (1979)

/12/ McMeekin, T.A., Patterson, J.T., Murray, J.G.: J. Appl. Bacteriol. 34, 699-716 (1971)

/13/ McMeekin, T.A., Shewan, J.M.: J. Appl. Bacteriol. 45, 321-332 (1978)

/14/ Priest, F.G., Somerville, H.J., Cole, J.A., Hough, J.S.: J. Gen. Microb. 75, 295-307 (1973)

/15/ Weeks, O.B.: p. 357-363 in Buchanan, R.E., Gibbons N.E. (ed.), Bergey's Manual of Determinative Bacteriology, 8th ed. Baltimore (1974): The Williams and Wilkins Co.

/16/ Yabuuchi, E., Tanimura, E., Ohyama, A., Yano, I., Yamamoto, A.: J. Gen. Appl. Microb. 25, 95-107 (1979)

# TAXONOMIC STUDIES ON HIGH G+C % FLAVOBACTERIA

P.R. Hayes, S.J. Mapp

Department of Microbiology, University of Leeds,
Leeds, LS2 9JT U.K.

## Summary

Using uv resistance and routine methods for preliminary screening, a total of 94 presumptive high GC % flavobacteria have been isolated from a wide variety of environments. These isolates, together with 8 high GC % reference cultures, form the strains being studied. Following detailed morphological, cultural, biochemical and nutritional studies four major groups have been provisionally delineated: group 1, oxidase positive and peritrichous (54 isolates); group 2, oxidase positive and non-motile (4 isolates); group 3, oxidase negative and peritrichous (7 isolates); group 4, oxidase negative and non-motile (29 isolates).

## Introduction

The genus Flavobacterium is divided into two sections in the 8th edn of Bergey's Manual of Determinative Bacteriology /7/. Six species are listed in each section which comprise, respectively, non-motile species with low G+C contents (30-45 %) and non-motile and peritrichous species with high G+C contents (53-69 %). Because of these large differences in G+C content it has been suggested that the high GC % species should be accommodated in a separate genus /2, 6/ and a more recently proposed emendation of the genus excludes these species /4/. Furthermore, one of the high GC % motile species, F. devorans, has been found to be motile by means of a single polar flagellum /8/, whilst two non-motile high GC % species, F. capsulatum and F. lutescens, have been shown to have affinities with other genera /1/.

Thus, there is now some doubt about the taxonomic status of the high G+C content species and this study is being undertaken in an attempt to clarify their validity and taxonomic position; in particular, the relationship between the non-motile and peritrichous forms is being assessed.

## Materials and Methods

### ISOLATION

McMeekin /5/ showed that it was possible to distinguish high G+C content flavobacteria from low G+C forms by using the greater resistance of the former to the detrimental effects of uv light. The efficacy of this technique was tested in the present study by using 15 'marker' flavobacteria of known G+C content (10 'high' and 5 'low'). The method involved taking 0.05 ml of diluted (ca. $10^3$ cells/ml) nutrient broth culture (24 h at $25^o$ C) which was spread in duplicate on to nutrient agar plates. One plate of each pair was exposed to a 15 W Hanovian uv lamp at a height of 58 cm for 90 sec. High G+C content strains showed a maximum reduction in numbers of ca. $10^2$ whilst low G+C content strains were reduced by ca. $10^4$ to $10^6$. The method was highly reproducible and a kill-off of $<5 \times 10^2$ was chosen as the basis for selecting high GC % isolates.

TABLE 1. Source of isolates

| Source | No. pigmented isolates | No. uv resistant | No. presumptive flavobacteria |
|---|---|---|---|
| Activated sludge | 181 | 75 | 21 |
| Water | 136 | 34 | 14 |
| Poultry | 111 | 33 | 16 |
| Beef | 76 | 31 | 4 |
| Pork | 59 | 15 | 11 |
| Leaf surfaces | 57 | 24 | 10 |
| Milk | 56 | 13 | 5 |
| Lamb | 27 | 5 | 4 |
| Soil | 27 | 1 | 1 |
| Fish | 9 | 4 | 2 |
| Frozen foods | 7 | 3 | 1 |
| Peat | 5 | 4 | 4 |
| Mushroom caps | 5 | 0 | 0 |
| Others | 3 | 1 | 1 |
| Total | 759 | 243 | 94 |

Many taxonomic studies including some one flavobacteria have suffered from a limitation in isolation environments. It is essential to examine a variety of sources, both to isolate a representative cross-section of strains and to determine any possible correlation between environment and strain characteristics. A total of 759 yellow pigmented Gram negative rods were isolated from the habitats listed (Table 1);

these cultures were screened by the uv method and 243 resistant isolates were obtained. However, since high GC % yellow pigmented Gram negative rods also include species in other genera including Pseudomonas, Xanthomonas and Erwinia, additional tests were necessary to delineate members of these genera. Those selected were Kovacs' oxidase test, method of attack on glucose (Hugh and Leifson test) and motility, with the position of the flagella determined /3/; confirmation of the flagella arrangement in representative isolates has yet to be performed. As a result of this secondary screening 94 out of 243 high G+C content isolates were retained as presumptive flavobacteria (Table 1). These isolates, together with 8 reference strains (Table 2) have formed the isolates being studied.

## TESTS INCLUDED FOR TAXONOMIC PURPOSES

A wide range of essentially routine morphological, cultural, environmental, biochemical and nutritional tests were performed. Included in the morphological and environmental tests were Gram stain, colonial appearance, gliding on Cytophaga agar, growth temperature range, growth in varying salt concentrations and antibiotic sensitivity patterns. Biochemical tests included dissimilation of a wide range of carbohydrated, hydrolysis of starch, casein and tributyrin, production of hydrogen sulphide, liquefaction of gelatin, $NO_3$ reduction, phosphatase and urease activity and changes on fresh blood agar.

For the nutritional studies the procedure adopted was that of Hayes /2/.

TABLE 2. Reference strains

| Nomenclature | Type culture collection number | G+C % |
|---|---|---|
| F. lutescens | Bu 2611 | 65 |
| F. devorans | NCIB 8195 | 69 |
| F. capsulatum | NCIB 980 | 63 |
| F. acidurans | ATCC 27383 | 66.4 |
| Flavobacterium sp. | NCIB 9942 | high |
| Flavobacterium sp. | NCIB 9491 | high |
| Flavobacterium sp. | NCMB 244 | 62.9 |
| Flavobacterium sp. | NCIB 11378 | 63.5 |

## NUMERICAL ANALYSIS

Test results for all isolates and reference strains are being coded at present, and similarities and cluster analyses will be performed, based on the University of Bradford Clustan 1A package. Unfortunately these results are not yet available.

## Results and Discussion

At the present time the isolates have been somewhat arbitrarily divided into 4 groups based on flagellation (non-flagellated or peritrichous) and oxidase reaction (+ or -) /3, 7/. The allocation of the 94 isolates to these 4 groups gave the following results: peritrichous, oxidase positive (54 isolates); non-motile, oxidase negative (29); peritrichous, oxidase negative (7); non-motile, oxidase positive (4). A correlation is seen between the source of isolates and their primary characters (Table 3). Thus oxidase positive, peritrichous isolates were readily obtained from activated sludge, water and milk whilst oxidase negative, non-motile isolates were mainly obtained from raw meats.

In terms of the overall results for the isolates, sensitivity to antibiotics was quite marked (between 60-80 % were sensitive to a wide range of antibiotics) and this contrasts with results obtained for low G+C content flavobacteria which are often relatively resistant. All isolates were catalase positive, all gave negative methyl red and Voges Proskauer reactions and none produced gas from carbohydrates.

Nutritionally, as with low GC % flavobacteria, the isolates were fairly diverse in their requirements. Thus 11 out of 94 grew in the inorganic salts + glucose medium, 40 required amino acids, 26 required growth factors and 17 failed to grow on any of the defined media used.

TABLE 3. Preliminary grouping of isolates and principal isolation sources

| Group allocation | No. of group isolations (total no. isolations) from source | | |
|---|---|---|---|
| Oxidase +, peritrichous | 18 | (21) | sludge |
| (54 isolates) | 10 | (14) | water |
| | 4 | (5) | milk |
| | 4 | (10) | leaf surfaces |
| Oxidase -, non-motile | 9 | (16) | poultry |
| (29 isolates) | 6 | (11) | pork |
| | 4 | (4) | lamb |

### PRELIMINARY GROUPING

Any attempt to group the organisms at this stage must be artificial and subject to error. However, the primary groups defined above have been more fully defined and subdivisions are suggested based on the test results obtained. These are:

Group 1. Oxidase positive, peritrichous (54 isolates). Three sub-groups distinguished: (a) comprises 22 isolates which are small rods, biochemically active, broadly antibiotic resistant and do not require growth factors; (b) 14 isolates, small rods to coccal forms, biochemically inert, antibiotic sensitive and require growth factors; (c) 11 isolates, medium rods, biochemically active, antibiotic sensitive and require growth factors.

Group 2. Oxidase positive, non-motile (4 isolates) - variable characteristics.

Group 3. Oxidase negative, peritrichous (7 isolates) - variable morphology, antibiotic sensitive, active biochemically and do not require growth factors.

Group 4. Oxidase negative, non-motile (29 isolates). Two sub-groups distinguished: (a) comprises 22 isolates which are short rods to coccal forms, relatively heat resistant and do not require growth factors; (b) comprises 3 isolates, medium rods, heat sensitive, antibiotic resistant, active biochemically and do not require growth factors.

Of the named reference strains F. lutescens and F. capsulatum would be included in Group 4 but the remaining reference cultures are phenetically atypical.

## CHARACTERS DISTINGUISHING HIGH AND LOW GC % FLAVOBACTERIA

The principal characters distinguishing high and low GC % flavobacteria are given in Table 4. At this rather early stage too much emphasis should not be placed on these features but it is interesting to note that the positive phosphatase reaction of many of the high GC % isolates is at variance with previously held views /3/.

TABLE 4. Differential characters of high and low GC % flavobacteria

| Test | % positive for test | |
|---|---|---|
| | High GC | Low GC[+] |
| Motility | 65 | 0 |
| Growth at $37^{o}$ | 48 | 10 |
| Heat resistance (50 C for 30 min) | 70 | 3 |
| Oxidase | 62 | 100 |
| Phosphatase | 50 | 100 |

[+]Results based on Hayes /2/

## References

/1/ Bauwens, M., De Ley, J.: this symposium

/2/ Hayes, P.R.: J. Appl. Bact. 43, 345-367 (1977)

/3/ Hayes, P.R., McMeekin, T.A., Shewan, J.M.: p. 177-187 in: Skinner, F.A., Lovelock, D.W. (eds), Identification Methods for Microbiologists, 2nd edn. London (1979): Academic Press.

/4/ Holmes, B., Owen, R.J.: Int. J. System. Bacteriol. 29, 416-426 (1979)

/5/ McMeekin, T.A.: J. Gen. Microbiol. 103, 149-151 (1977)

/6/ McMeekin, T.A., Shewan, J.M.: J. Appl. Bact. 45, 321-332 (1978)

/7/ Weeks, O.B.: p. 357-364 in: Buchanan, R.E., Gibbons, N.E. (eds), Bergey's Manual of Determinative Bacteriology, 8th edn. Baltimore (1974): Williams & Wilkins Co.

/8/ Yabuuchi, E., Tanimura, E., Ohyama, A., Yano, I., Yamamoto, A.: Abst. 12th Int. Congr. Microbiol. Munich, p. 90 (1978)

# IDENTIFICATION AND CLASSIFICATION OF *FLAVOBACTERIUM* SPECIES FROM CLINICAL SOURCES

R. J. Owen, B. Holmes

National Collection of Type Cultures, Central Public Health Laboratory, Colindale Avenue, London NW9 5HT, UK

## Summary

Clinical isolates of aerobic yellow-pigmented, Gram-negative bacteria allied to *Flavobacterium* are classified in five taxa of which *F. meningosepticum*, *F. odoratum* and Group IIb are of some importance as human pathogens. Antibiotic resistance is a feature of these bacteria and preliminary data suggest their resistance is chromosomally determined and not plasmid mediated. The taxa can be separated from each other by various biochemical tests and by data on chromosomal DNA characteristics. The similarities within and between species of *Flavobacterium* can be conveniently illustrated by means of bivariate scatter diagrams, with G+C content and base pairing as the variables. It is evident from such diagrams that most clinical species of *Flavobacterium* and unnamed taxa are comprised of a 'core' of highly related strains (typical strains) with a number of 'outliers' (atypical strains). The classification of the latter strains is problematic since they are often indistinguishable from typical strains in conventional phenotypic tests. The food and environmental strains so far examined are often atypical when compared with the clinical isolates.

## Introduction

In 1965, a service was initiated to assist clinical laboratories in the UK with the identification of so-called 'difficult' bacteria for which no reference laboratories existed. The experience gained since then has shown that yellow-pigmented, rod-shaped, Gram-negative bacteria, in particular those allied to the genus *Flavobacterium* constitute one of several problem groups encountered in the clinical laboratory. *Flavobacterium meningosepticum* is well known to medical microbiologists in many parts of the world [27], but until quite recently the identification of other species of *Flavobacterium* was almost impossible. This was due mainly to the lack of suitable reference strains, poor species descriptions, and the unsuitability of many conventional biochemical tests for the characterization of clinical isolates.

Improvements to the probability matrix used in the NCTC for computer-assisted identification of Gram-negative bacteria were hindered by the taxonomic confusion which has been a feature for many years of the genus Flavobacterium [10, 14]. Even so, F. meningosepticum has provided a focal point for medical microbiologists, and initial identifications usually were based on comparisons with that species. During the last decade, various groups of Flavobacterium-like bacteria from clinical sources have been studied, and now at least five major taxa are recognised. Flavobacteria are not part of the normal human flora, but they are of increasing interest to medical microbiologists because of their involvement in hospital acquired infections [9] and their unusually high resistance to the antibiotics commonly used to treat Gram-negative infections [12, 28]. The role of environmental, in particular water-borne, bacteria in human infection is highly topical as shown by the recent work on Legionella pneumophila and allied bacteria [1].

The purpose of this presentation is to describe the methods used in the NCTC to identify Flavobacterium strains from clinical sources. These are based on various conventional biochemical and morphological characteristics routinely used in medical laboratories, and on the characterization of chromosomal DNA. Recently, some strains were examined for the presence of plasmid DNA and preliminary data on this aspect of Flavobacterium will be described.

## The main taxa

The five taxa (three named species and two unnamed groups) to which isolates of Flavobacterium from clinical and other sources of public health interest can be classified are listed in Table 1. A further unnamed group, Group IIf, is included in the Table and elsewhere in the text as this group is often referred to in the literature as Flavobacterium-like although its precise taxonomic position remains undetermined [20]. The pathogenic importance of these bacteria has been dealt with elsewhere [27], but it should be emphasized that although F. meningosepticum is the accepted pathogen, there is evidence that taxa such as F. odoratum and Group IIb are also of importance. The CDC Group IIb strains are in our experience the most common

TABLE 1

The Flavobacterium taxa encountered in medical microbiology

| Taxon | Comment |
|---|---|
| F. breve | Non-pathogen |
| F. meningosepticum | Pathogen [13, 27] |
| F. odoratum | Occasional pathogen [2] |
| CDC Group IIb | Occasional pathogen [27] |
| CDC Group IIk, type 2 | Occasional pathogen [4] |
| CDC Group IIf | Non-pathogen |

of the Flavobacterium isolates from clinical specimens and the hospital environment in the UK. They occur in a variety of specimens and occasionally have been incriminated in infections of patients; notably as the cause of localised abscesses and pus lesions.

Members of these Flavobacterium taxa have in common the characteristics shown in Table 2.

TABLE 2

Identification of clinical Flavobacterium isolates

| | |
|---|---|
| Colonies: | on nutrient media typically circular and smooth; pigmented yellow to orange. Aerobic. |
| Morphology: | straight rods (2-3μ). Gram-negative. Non-sporing. |
| Oxidase + Catalase + Phosphatase + O-F test oxidative or no change | |
| Optimum growth temperature 30 to 37°C | |
| Generally actively proteolytic, and weakly saccharolytic | |

## Taxon identification

### Overall phenetic similarity

Similarities between the taxa mentioned above except Group IIf were recently investigated by numerical taxonomic methods [Holmes and Owen, unpublished results]. The inter-relationships between representative strains of the various taxa based on a comprehensive range of phenotypic tests are illustrated in Fig. 1. The purpose of the numerical taxonomy was to investigate the affinities of Group IIk, type 2, consequently the other Flavobacterium taxa were represented only by a small number of strains. Even so the relative order of their similarities is demonstrated by the single linkage cluster analysis, as is the heterogeneity of Group IIb and its close

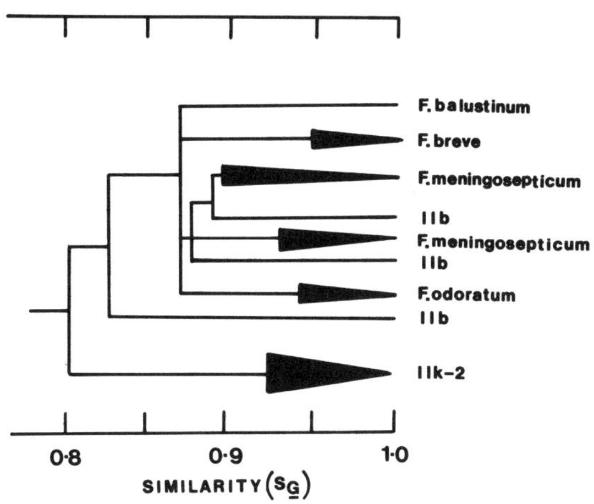


Fig. 1. Simplified dendrogram based on single-linkage cluster analysis. Similarity was calculated with the general similarity coefficient of Gower ($S_G$), negative matches included.

affinities to F. meningosepticum. It was concluded from this study that CDC Group IIk, type 2 should be included in the genus Flavobacterium and that it should represent a new species. We have proposed that it should be called F. multivorum [Holmes, Owen and Weaver, unpublished results].

Test differences

The morphological and biochemical test differences useful for distinguishing between the Flavobacterium taxa are listed in Tables 3 and 4. The taxa for convenience are divided into those that are saccharolytic, that is generally produce acid from glucose and maltose (Table 3), and those that are non-saccharolytic (Table 4). A similar range of distinguishing tests was given by Gilardi [5].

TABLE 3

Identification of Flavobacterium : saccharolytic taxa*

| Test | F. breve | F. meningosepticum | Group IIb | Group IIk, type 2 |
|---|---|---|---|---|
| Acid, trehalose | - | + | + | + |
| Acid, sucrose | - | - | - | + |
| Acid, salicin | - | - | - | + |
| Acid, raffinose | - | - | - | + |
| Growth at 42°C | - | 3/9† | 2/4 | - |
| Nitrite reduction | - | 1/9 | 1/4 | - |
| Starch hydrolysis | - | - | + | - |
| Urease production | - | - | - | 27/28 |
| Indole production | + | + | + | - |
| Yellow pigment | + | 5/9 | + | + |
| Pigment on tyrosine | - | 4/9 | 3/4 | 1/28 |
| β-galactosidase | - | + | 1/4 | - |
| Casein digestion | + | + | + | - |
| Aesculin hydrolysis | - | + | + | + |

* Most strains produce acid from glucose and maltose
† Number of positive/number tested

The taxa in Table 3 have several important features. F. meningosepticum and Group IIb are biochemically similar although not identical because the latter is more heterogeneous in biochemical characteristics. Their main consistent difference is in hue of pigmentation (F. meningosepticum is weakly yellow, Group IIb is bright yellow). Group IIk, type 2 is unusual among the Flavobacterium taxa in producing acid from a wide range of carbohydrates. It also differs from F. meningosepticum and the other saccharolytic species in neither producing indole nor digesting casein. F. breve resembles F. meningosepticum except it does not produce acid from glycerol or from trehalose, and it fails to hydrolyse aesculin, and does not produce β-galactosidase.

TABLE 4

Identification of Flavobacterium: non-saccharolytic taxa

| Test | F. odoratum | Group IIf |
|---|---|---|
| Growth at 42°C | - | + |
| Nitrite reduction | + | - |
| Starch hydrolysis | - | 1/16 |
| Urease production | + | - |
| Indole production | - | + |
| Yellow pigment | 27/28 | - |
| Pigment on tyrosine | 3/28 | + |
| Tyrosine hydrolysis | 13/28 | - |
| β-galactosidase | - | - |
| Casein digestion | + | + |
| Penicillin sensitivity | - | + |

In Table 4, F. odoratum is characteristically indole-negative and non-saccharolytic. It differs from most other Flavobacterium strains in reducing nitrite and in producing urease. Group IIf is also non-saccharolytic but is indole-positive. The group is not pigmented on nutrient agar, and its mucoid colonies are distinctive and are unlikely to be confused with those of F. odoratum, which are yellow-pigmented and non-mucoid.

Serology

Because F. meningosepticum was incriminated in several hospital outbreaks of neonatal meningitis, a serological scheme was developed in the 1950's to assist epidemiological surveillance [13]. Six serological groups (A to F) of F. meningosepticum were recognised and most of the strains from hospital outbreaks belonged to serotype C. Recently in France, Richard and co-workers [23,24] described a further six serotypes (G to L). These strains were mainly from blood cultures and serotype G was the commonest. In our identification scheme only strains representing serotypes G, H and K, were confirmed as F. meningosepticum. The remaining three strains appeared to have closer affinities to F. breve or were not identified.

The serology of other species of Flavobacterium has not been developed to any extent although some serological cross-reactions between strains of Group IIb and F. meningosepticum, and Group IIf and F. meningosepticum have been reported [22]. Serology is therefore of limited value at present in the basic identification of flavobacteria from clinical sources.

## Characteristics of chromosomal DNA

For the past 15 years, investigations of base composition, genome size and base sequence similarity have made a significant contribution to bacterial taxonomy [17]. The techniques are highly specialized and only appropriate for a reference laboratory concerned with the identification of problem groups. It is in this context that DNA characterization has proved of considerable value in the investigation of flavobacteria.

Base contents

In 1963, De Ley and van Muylem [3] determined the G+C contents of various flavobacteria from marine environments. It was shown that several of these strains, which were believed to be like Flavobacterium or Cytophaga in conventional characteristics [7], were significantly different in the base contents of their DNAs. Subsequent DNA investigations of named species of Flavobacterium confirmed their heterogeneity with values ranging from 30 to 70% G+C.

In 1974, Weeks [29] divided Flavobacterium into two sections. Section I contained species with G+C contents of 30 to 42% (low G+C content), whereas Section II contained the high G+C content strains (>42%). Although restriction of Flavobacterium to include only certain low G+C content species has been proposed [10], the publication in 1980 of the Approved Lists of Bacterial Names [25] has re-introduced confusion because various species with high G+C contents were retained in the genus as well as some Gram-positive taxa considered species incertae sedis by Weeks [29].

Table 5 lists the base compositions of the six taxa of medical importance in or allied to the genus Flavobacterium as defined by Holmes and Owen [10]. All strains have values in the 31 to 41% G+C range. F. odoratum is unusual as it comprises two

TABLE 5

G+C contents of Flavobacterium, medical taxa

| Taxon | G+C content (mol %) | |
|---|---|---|
| | range | mean ± SD of typical strains |
| F. breve (10 strains) | 31.2 - 33.1 | 32.4 ± 0.6 |
| F. meningosepticum (8 strains) | 36.1 - 37.6 | 37.0 ± 0.5 |
| F. odoratum (10 strains) | 31.4 - 36.1 | 31.9 ± 0.4 |
| Group IIb (16 strains) | 35.0 - 38.9 | 37.4 ± 1.3 |
| Group IIk, type 2 (11 strains) | 38.5 - 40.1 | 39.5 ± 0.5 |
| Group IIf (6 strains) | 37.0 - 38.0 | 37.4 ± 0.4 |

distinct DNA subgroups of 31 to 33% and 34 to 36% G+C respectively [18]. The most common strains in clinical specimens belong to the former subgroup, but the type strain itself [11] belongs to the latter group. In conventional biochemical characteristics, however, there is no indication of a significant phylogenetic division within the species. Group IIb also has a relatively broad range of G+C contents, but this heterogeneity is reflected also at the morphological and biochemical level.

Genome size

The size of the double-stranded DNA molecule that comprises the bacterial genome is usually expressed in molecular weight units (daltons). Table 6 lists estimates of genome size for the various Flavobacterium taxa of medical importance. These estimates

TABLE 6

Genome sizes of Flavobacterium,
medical taxa excluding Group IIk, type 2

| Taxon | genome size (daltons) | |
|---|---|---|
| | range | mean ± SD x $10^{-9}$ |
| F. breve (10 strains) | 2.72 - 3.85 | 3.21 ± 0.37 |
| F. meningosepticum (3 strains) | 2.50 - 2.85 | 2.73 ± 0.20 |
| F. odoratum (10 strains) | 2.52 - 3.92 | 2.90 ± 0.46 |
| Group IIb (3 strains) | 2.85 - 3.52 | 3.15 ± 0.34 |
| Group IIf (2 strains) | 1.64 - 1.81 | 1.73 ± 0.12 |

were obtained from the spectrophotometric determination of initial renaturation rates. Except for Group IIf, representative strains of the taxa had genome sizes in the 2.5 to 3.9 x $10^9$ dalton range. From available data, this range is typical of other bacterial genera [17]. Group IIf has a genome approximately 40% smaller than those of typical Flavobacterium species which further supports its exclusion from the genus. As Group IIf has some phenotypic similarities to Moraxella, it is interesting that its small genome size compares closely to the size range of the species of that genus.

Base sequence similarities

In 1973, Sottile and co-workers [26] reassociated DNA from various F. meningosepticum strains using DNA from the type strain (King 14, NCTC 10016) as the reference. Their results indicated high levels of base sequence divergence within the species. A subsequent investigation in our laboratory [21] confirmed that the type strain was indeed atypical, and it shared on average only 30% of its DNA

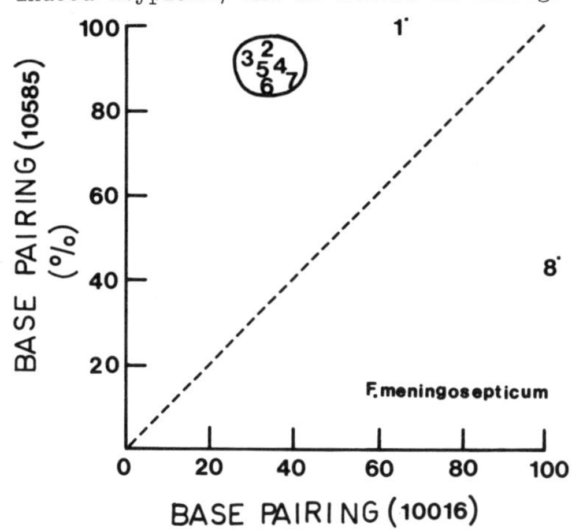


Fig. 2. Reassociation with F. meningosepticum: 1, NCTC 10585; 2, NCTC 10587; 3, NCTC 10589; 4, F134; 5, NCTC 10586; 6, NCTC 10588, 7, F123; 8, NCTC 10016.

sequences with most of the other F. meningosepticum strains, including strains of the same serotype. However, another reference DNA (King 422, NCTC 10585) was found to be highly related to all the F. meningosepticum strains except for strain 14, thus confirming that the type strain was atypical. The results on F. meningosepticum are summarized in Fig. 2. The high degree of similarity between most of the F. meningosepticum DNAs was supported by results on thermal stabilities of DNA duplexes [21].

The results on F. meningosepticum indicated the importance of selecting suitable DNA references. Misinterpretations of results can be avoided if several reference DNAs are used and if results are assessed in conjunction with conventional test results and other molecular data. The relationships between F. meningosepticum and representatives of the various Flavobacterium taxa were investigated using reference DNA from King strain 422, and these are illustrated in Fig. 3. To assist in understanding the inter-relationships of strains, the results are presented in the form of bivariate scatter diagrams. The first variable, G+C content, is fixed for each strain, but the second variable, base pairing, is dependent on the reference strain used. Fig. 3 shows that F. meningosepticum can be clearly separated from other Flavobacterium taxa on its DNA characteristics.

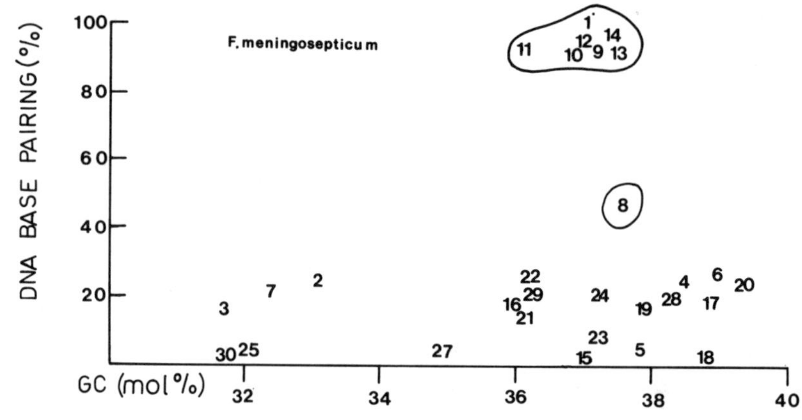


Fig. 3. Bivariate scatter diagram of DNA similarities between F. meningosepticum NCTC 10585 reference DNA (1)and allied bacteria. Serial numbers: 1, 8 through 14, F. meningosepticum; 2, F. balustinum; 3, 29, F. odoratum; 4, 15 through 22, Group IIb; 6, Group IIk, 2; 5, 23, 24, Group IIf; 25, F. aquatile; 27, F. pectinovorum; 28, Moraxella saccharolytica; 7, 30, F. breve.

## Flavobacterium Group IIb

The base sequence relationships among strains within this group were investigated because of their close resemblance in phenotypic characteristics and base compositions to F. meningosepticum. Group IIb strain NCTC 10795, selected as the DNA reference, was representative of the taxon in conventional phenotypic characteristics. The bivariate scatter diagram for DNA of this taxon and allied flavobacteria is given in Fig. 4. The results indicated significant levels of heterogeneity within Group IIb and low levels of base sequence similarity to F. meningosepticum and allied taxa.

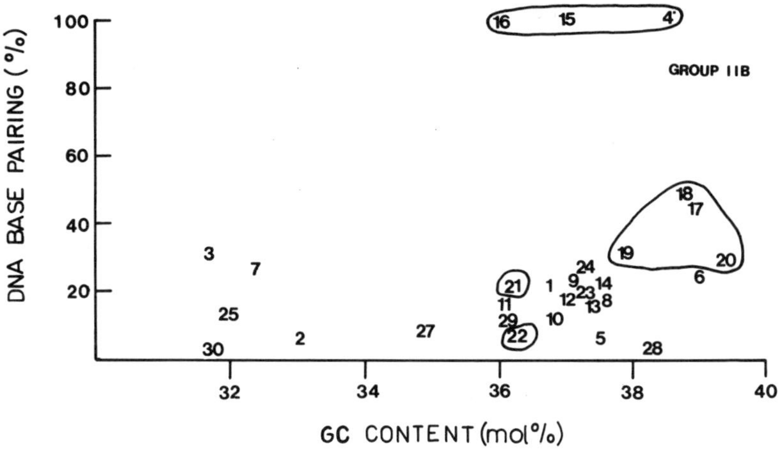


Fig. 4. Bivariate scatter diagram of DNA similarities between Group IIb NCTC 10795 reference DNA and allied bacteria. See Fig. 3 legend for serial numbers.

In 1978, Brenner and co-workers [1] investigated DNA relatedness between Legionella pneumophila and strains of F. meningosepticum and Group IIb. It was concluded that Legionella was not related to these organisms at the species or genus level, even though they had similarities in G+C content, fatty-acid contents, pigmentation and antibiotic resistance profiles.

Other Flavobacterium taxa

Similar analyses of Flavobacterium were made using reference DNA from F. odoratum and F. breve. Fig. 5 illustrates the results obtained for F. odoratum which contains at least 2 DNA groups [18]. The results have been described elsewhere [19] for F. breve and again some heterogeneity within the species was found.

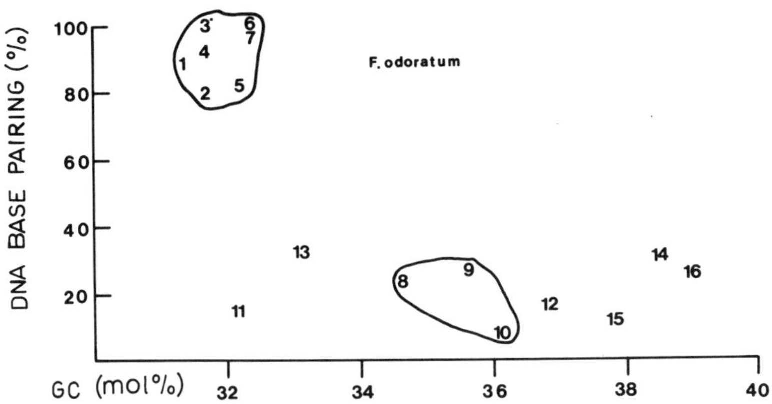


Fig. 5. Bivariate scatter diagram of DNA similarities between F. odoratum CL607/74 reference DNA and allied bacteria. Serial numbers: 1 through 10, F. odoratum; 11,; F. breve; 12, F. meningosepticum; 13, F. balustinum; 14, Group IIb; 15, Group IIf; 16, Group IIk, type 2.

In general, the results on F. breve, F. meningosepticum and F. odoratum indicated that these species were comprised of a 'core' of highly related strains. The majority of strains are typical in that they belonged to this 'core', but there are a number of 'outliers' or atypical strains, with varying degrees of similarity to the typical strains. At present, only limited reassociation data are available on the unnamed groups IIk, type 2 and IIf. Results with reference strains from other taxa indicated the groups do not have significant levels of DNA similarity to any of the named species.

## Diversity among species

The overall diversity in DNA base sequence relatedness between the various Flavobacterium taxa from clinical sources is illustrated in Fig. 6. This DNA similarity matrix (7 x 7) was compiled from reassociation results on reference strains representing each taxon. Thus, the strains with the exception of Group IIf, all shared on average 23% of their base sequences as determined under optimal renaturation conditions. The levels of divergence within their shared sequences may differ from strain to strain; even so, the taxa had sufficient DNA similarities to be considered members of the same genus. However, Group IIf had on average only 12% of its DNA sequences in common with the other taxa and this further supports the view that the group belongs in a separate genus from Flavobacterium.

DNA SIMILARITY MATRIX (%)

| | | A | B | C | D | E | F | G |
|---|---|---|---|---|---|---|---|---|
| F. balustinum | A | 100 | | | | | | |
| F. odoratum | B | 32 | 100 | | | | | |
| F. breve | C | 24 | 22 | 100 | | | | |
| Group IIk-2 | D | 30 | 25 | 22 | 100 | | | |
| F. meningosepticum | E | 24 | 16 | 21 | 26 | 100 | | |
| Group IIb | F | 5 | 30 | 27 | 23 | 21 | 100 | |
| Group IIf | G | 15 | 16 | 6 | 25 | 4 | 6 | 100 |

Fig. 6. Similarity matrix of Flavobacterium taxa derived from DNA reassociation data.

## Plasmid DNA

Extra-chromosomal DNA elements such as plasmids are widespread in bacteria although their presence in Flavobacterium has not been previously reported. The fact that the clinical isolates are often highly resistant to antibiotics suggested the possibility that their resistance was plasmid mediated. There are not as yet established systems in flavobacteria for investigation of classical genetic phenomena such as transformation and transduction. However, a preliminary study has been made of selected strains to determine if a broad correlation exists between the presence of plasmids and antibiotic resistance of these bacteria. Representatives of each of the main taxa of Flavobacterium were screened for plasmids by means of the single colony lysis technique. The total of 22 strains examined included strains of F. breve, F. meningosepticum and F. odoratum which were resistant to therapeutic levels of commonly used antibiotics such as gentamicin, streptomycin and tetracycline. Plasmid DNA was not detected in typical resistant strains belonging to any of those species. It seems likely therefore that the antibiotic resistance characteristic of the clinical isolates of Flavobacterium, is not due to R factors and that chromosomal genes probably play a significant role in antibiotic response.

## Conclusion

There are several problems yet to be resolved in the classification of the Flavobacterium-Cytophaga complex and two aspects in particular are pertinent to the taxa from clinical sources.

(a) Atypical and unclassified clinical strains. The taxa described above provide a useful basis for the initial identification of clinical isolates of flavobacteria. However, various strains do not fit readily into any of the recognised taxa in Flavobacterium. The identification and classification of these strains is particularly problematic when they are indistinguishable from typical strains in conventional phenotypic tests yet differ in DNA characteristics. At present the unidentifiable strains are designated according to the taxon they resemble most closely; so if they resemble Group IIb they are referred to as 'Group IIb-like'. These strains may eventually form the basis of new taxa.

(b) Food and environmental strains. Strains from non-medical sources have been included in our studies wherever possible. It was noticeable that they commonly could not be identified. A comparison was made between our taxa and those described in the comprehensive studies of Hayes and co-workers [8]. The results showed that Hayes's phenon 1 corresponded closely to Group IIb; in addition, Hayes's phenon 2 and phenon 4 were confirmed as similar to F. breve, although the latter phenon was atypical in G+C content. F. meningosepticum has been included as a reference species in several numerical taxonomic studies of the flavobacteria [6, 15], but it failed to cluster with any environmental strains. F. meningosepticum is probably rare in natural habitats although Olsen [16] described isolates from river water.

## Acknowledgements

We thank Marilyn Nugent for screening strains for the presence of plasmid DNA.

## References


[1] Brenner, D.J., Steigerwalt, A.G., Weaver, R.E., McDade, J.E., Feeley, J.C., Mandel, M. : Current Microbiol. 1, 71-75 (1978)

[2] Davis, J.M., Peel, M.M., Gillians, J.A. : Med. J. Aust. 2, 703-704 (1979)

[3] De Ley, J., Van Muylem, J. : Ant. van Leeuw. J. Microbiol. Serol. 29, 344-358 (1963)

[4] Dhawan, V.K., Rajashekariaiah, K.R., Metzger, W.I., Rice, T.W., Kallick, C.A. : J. Clin. Microbiol. 11, 492-495 (1980)

[5] Gilardi, G.L. : p. 45-65 in : Gilardi, G.L. (ed.), Glucose non-fermenting gram-negative bacteria in clinical microbiology. Florida (1978) : CRC Press, Inc.

[6] Goodfellow, M., Austin, B., Dickinson, C.H. : J. Gen. Microbiol. 97, 219-233 (1976)

[7] Hayes, P.R. : J. Gen. Microbiol. 30, 1-19 (1963)

[8] Hayes, P.R., Wilcock, A.P.D., Parish, J.H. : J. Appl. Bacteriol. 43, 111-115 (1977)

[9] Herman, L.G. : Adv. Appl. Microbiol. 23, 155-171 (1978)

[10] Holmes, B., Owen, R.J. : Int. J. Syst. Bacteriol. 29, 416-426 (1979)

[11] Holmes, B., Snell, J.J.S., Lapage, S.P. : Int. J. Syst. Bacteriol. 27, 330-336 (1977)

[12] Holmes, B., Snell, J.J.S., Lapage, S.P. : J. Clin. Pathol. 32, 73-77 (1979)

[13] King, E.O. : Am. J. Clin. Pathol. 31, 241-247 (1959)

[14] McMeekin, T.A., Shewan, J.M. : J. Appl. Bacteriol. 45, 321-332 (1978)

[15] McMeekin, T.A., Stewart, D.B., Murray, J.G. : J. Appl. Bacteriol. 35, 129-137 (1972)

[16] Olsen, H. : Acta Pathol. Microbiol. Scand. 75, 313-322 (1969)

[17] Owen, R.J., Hill, L.R. : p. 277-296 in : Skinner, F.A., Lovelock, D.W. (eds), Identification methods for microbiologists.SAB Technical Series No. 14. London (1979) : Academic Press

[18] Owen, R.J., Holmes, B. : FEMS Microbiol. Lett. 4, 41-46 (1978)

[19] Owen, R.J., Holmes, B. : Current Microbiol. 4, In press (1980)

[20] Owen, R.J., Snell, J.J.S. : Ant. van Leeuw. J. Microbiol. Serol. 39, 473-480 (1973)

[21] Owen, R.J., Snell, J.J.S. : J. Gen. Microbiol. 93, 89-102 (1976)

[22] Price, K.W., Pickett, M.J. : this Symposium

[23] Richard, C., Monteil, H., Laurent, B. : Ann. Microbiol. (Inst. Pasteur) 130B, 141-144 (1979)

[24] Richard, C., Monteil, H., Le Faou, A., Laurent, B. : Méd.Mal. Infect. 9, 124-128 (1979)

[25] Skerman, V.B.D., McGowan, V., Sneath, P.H.A. (eds) : Int. J. Syst. Bacteriol. 30, 225-420 (1980)

[26] Sottile, M.I., Baldwin, J.N., Weaver, R.E. : Appl. Microbiol. 26, 535-539 (1973)

[27] von Graevenitz, A. : p. 119-153 in : Gilardi, G.L. (ed.), Glucose non-fermenting gram-negative bacteria in clinical microbiology. Florida (1978) CRC Press, Inc.

[28] von Graevenitz, A., Grehn, M. : FEMS Microbiol. Lett. 2, 289-292 (1977)

[29] Weeks, O.B. : p. 357-364 in : Buchanan, R.E., Gibbons, N.E. (eds), Bergey's Manual of Determinative Bacteriology. Baltimore (1974) Williams and Wilkins

# YELLOW-PIGMENTED NONFERMENTATIVE BACILLI: TAXONOMY OF THE IIk-2 GROUP

M. J. Pickett , M.G. Levine , M. Mandel

Department of Microbiology, University of California, Los Angeles, CA 90024, USA, and Department of Molecular Biology, M.D. Anderson Hospital and Tumor Institute, The University of Texas System Cancer Center, Texas Cancer Center, Houston, Texas 77030, USA

## Summary

Yellow-pigmented nonfermentative bacilli recovered from clinical specimens were grouped according to their salient features by Elizabeth King. She first assigned numbers to these groups. Many of these are now well characterized and have been named. This report reviews these yellow-pigmented bacilli, presents our taxonomic studies on the IIk-2 group, and concludes that it should be assigned to the genus *Flavobacterium*.

The more commonly encountered, yellow- to orange-pigmented, nonfermentative bacilli such as the flavobacteria and *P.stutzeri* have been described by both taxonomists and clinical bacteriologists for several decades. Other yellow, nonfermentative bacilli (YNF) are only rarely recovered from clinical specimens and only recently have been intensively studied in respect to their features, taxonomy, and suitable names.

This report reviews, particularly from the point of view of clinical bacteriology, developing knowledge of the YNF, and presents our own studies on the "IIk" group with emphasis on one biovar, IIk-2. A convenient starting point for this review is the 1964 report of Elizabeth King from the Communicable Disease Center ("CDC", now Center for Disease Control, Atlanta, Georgia, USA). She presented, at a "Roundtable" during the annual meeting of the Society of American Bacteriologists, her observations on clinical isolates including nonfermentative bacilli. Among the groups of bacteria she recognized were several which presented yellow to orange, nondiffusible pigment on nutrient agar medium. At that time she tabulated four groups of YNF (Table 1). These were two flavobacteria (IIa and IIb), Pseudomonas stutzeri (Vb), and her "eugonic oxidizer group 1" (EO-1). She noted that the two saccharolytic groups of flavobacteria were indole positive and strongly proteolytic. Group IIa, already at that time addressed as Flavobacterium meningosepticum, acidified lactose and mannitol in a nutrient broth basal medium. The P.stutzeri group differed from the others in that it produced gas from nitrate. The EO-1 group was later equated with Pseudomonas multivorans of Stanier, Palleroni, and Doudoroff /24/, and still later with the plant pathogen, Pseudomonas cepacia.

Table 1. Yellow-pigmented nonfermentative rods*

| | Flavobacterium | | P.stutzeri | |
|---|---|---|---|---|
| | IIa | IIb | Vb | EO-1 |
| Oxidase | + | + | + | -/+w |
| Acid from (NBBM)** | | | | |
| lactose | + | - | - | +/- |
| mannitol | + | - | - | - |
| Indole | +w | +w | - | - |
| Nitrate → nitrite | - | - | + | -/+ |
| Nitrate → gas | - | - | + | - |
| Milk, peptonized | + | + | - | +/- |
| Motility | - | - | + | +/- |

*Adapted from King /12/.

**NBBM, nutrient broth basal medium; w, weak; +/-, usually positive.

Elizabeth King's Roundtable has undergone several revisions. The 1972 revision by Weaver, Tatum, and Hollis introduced four more groups of YNF (Table 2). The two groups of flavobacteria remained, but were no longer readily distinguished in tests for acidification of lactose and mannitol. However, the intensity of their pigmentation provided tentative differentiation. The four new groups differed from the flavobacteria in being indole negative, and they were not strongly proteolytic. Weaver et al. distinguished the IIk and Ve groups by oxidase, lactose, and mannitol tests. The two biovars within each of these groups were not sharply distinguished in their tables.

Table 2. Yellow-pigmented nonfermentative rods*

| | Flavobacterium | | "P.stutzeri" | | | | |
|---|---|---|---|---|---|---|---|
| | IIa | IIb | Vb | IIk-1 | IIk-2 | Ve-1 | Ve-2 |
| Oxidase | + | + | + | V** | + | - | - |
| Acid from (OF) | | | | | | | |
| lactose | -/+ | - | - | + | + | - | - |
| mannitol | +/- | -/+ | +w | - | - | + | + |
| Indole | + | + | - | - | - | - | - |
| Milk, peptonized | + | + | - | - | - | -/+ | - |
| Nitrate → gas | - | - | + | - | - | - | - |
| Pigment | +w | +s | +w | +/- | -/+w/+ | + | + |

*Adapted from Weaver et al. /26/.
**V, strain variation; OF, CDC's OF basal medium; -/+, usually negative; w, weak; s, strong.

In retrospect, our understanding of the YNF was momentarily confused rather than clarified during the past decade by Pickett & Pedersen /21/, and later by Gilardi /3/. On grounds of pigmentation and biochemical features, they addressed their clinical isolates of IIk as Xanthomonas. Our knowledge of the IIk group was summarized in 1974 by Tatum, Ewing, and Weaver /25/ as follows:

> "...Members of two groups (IIk and Ve) produce yellow pigment. At present, the taxonomic position of these bacteria is undecided, but their characteristics are such that they may be considered to be members of the genus Pseudomonas, or possibly of the genus Xanthomonas."

Certainly, as discussed below and noted by Gilardi in 1975 /3/, there are marked similarities between xanthomonads and strongly pigmented strains of IIk. Indeed, in his 1975 report Gilardi stated that "...it has been proposed that IIk and xanthomonads be placed in the genus Pseudomonas as a single genetic species, P.campestris."

The 1972 tables of Weaver and his associates /26/ did not appear to provide a comfortable distinction between IIk-1 and IIk-2. However, this distinction was sharpened in the tables they presented /25/ in 1974 (Table 3). Note in this table that both pigmentation and the urea test serve to distinguish these two biovars. Parenthetically we would note here that other workers also have found the urease test to be nighly reliable for distinguishing these two biovars.

Table 3. IIk: Differential features*

| Biovar: | IIk-1 | IIk-2 |
|---|---|---|
| Number of strains: | 100 | 40 |
| Pigment (yellow) | +s | - or +w |
| | %+ | |
| Acid from | | |
| inulin | 12 | 90 |
| starch | 0 | 88 |
| Growth on MacConkey agar | 23 | 100 |
| Oxidase | 75 | 100 |
| Urease | 4w | 100 |

*Adapted from Tatum, Ewing, & Weaver /25/.

The report of Holmes, Owen, and their associates in 1977 /7/ marked a major advance in clarifying the taxonomy of the IIk group. They found (Table 4) that the chromosomal DNA of IIk-2 contained only 40 moles% guanine+cytosine. IIk-2, therefore, was excluded from consideration in the genus Pseudomonas /15/. In their report they proposed the name Pseudomonas paucimobilis for CDC's IIk-1 group. Note in this table that (1) all of the three taxa (8 strains of xanthomonads, 29 of P.paucimobilis, and 2 of IIk-2) acidified sucrose, and (2) several features served to distinguish xanthomonads and P.paucimobilis. Unfortunately, some of these distinguishing features have not been found by other workers. In particular, "growth on MacConkey agar" is scored differently by Gilardi /4/, Tatum et al. /25/, and Levine /13/. We suggest that size of inocula, and both temperature and duration of incubation may account for some of these differences.

In our studies on the Xanthomonas-IIk complex*, we too looked for features that would distinguish the three groups of Table 4. Using 9 strains of xanthomonads, 19 of P.paucimobilis, and 22 of IIk-2, we obtained the results shown in Table 5. All of our xanthomonads, but none of our strains of IIk-1, presented mucoid colonies. Holmes and his associates reported that 28 of their 29 strains of P.paucimobilis presented smooth colonies, and their one mucoid strain was also aberrant in several other features. They also distinguished Xanthomonas and P.paucimobilis by tests for acidification of ethanol, raffinose, and salicin (Table 4), but these tests did not clearly distinguish our strains (Table 5). Note finally in Table 5 that we too found pigmentation, acidification of ethanol, and alkalinization of urea to be practical and useful features for distinguishing strains of IIk-1 and IIk-2.

*Details of our procedures are given in Levine /13/ and Levine et al. /14/.

Table 4. Features of yellow-pigmented IIk and *Xanthomonas**

| | *X. campestris* | *P. paucimobilis* (IIk-1) | IIk-2 |
|---|---|---|---|
| Moles% G+C | 68.2 | 69.9±1.4 | 39.6 |
| Acid from | | | |
| ethanol | - | + | - |
| mannitol | - | - | - |
| raffinose | - | + | + |
| salicin | - | + | + |
| sucrose | + | + | + |
| Growth, MacConkey agar | + | - | + |
| Motility | + | + | - |
| PBHB granules** | - | + | - |

*Adapted from Holmes et al. /7/.

**PBHB, Poly-beta-hydroxybutyrate; +, 85% or more of strains positive; -, 15% or less positive.

Table 5. Differential features of IIk and *Xanthomonas*

| | *Xanthomonas* | IIk-1 | IIk-2 |
|---|---|---|---|
| Moles% G+C | 65.3 | 65-66 | 41-45 |
| Colonies, 24-48 hr | Mucoid, yellow | Smooth, yellow | Smooth colorless |
| | | %+ | |
| Acid from | | | |
| ethanol | 11 | 95 | 0 |
| inulin | 0 | 16 | 100 |
| lactose | 11 | 100 | 100 |
| raffinose | 89 | 95 | 100 |
| salicin | 11 | 100 | 100 |
| Growth, MacConkey agar | 22 | 0 | 9 |
| Motility, soft agar | 0 | 21 | 0 |
| Urease | 0 | 0 | 100 |

Table 6. Xanthomonas and P.paucimobilis: Summary data

| | Xanthomonas | P.paucimobilis |
|---|---|---|
| Pathogenic for: | plants | man |
| Moles% G+C | 65-68 | 65-66 |
| Pigment | xanthomonadin | nostoxanthin |
| Colonies | mucoid | smooth |
| Cellular fatty acid group | 6? | 9 |
| Acid from | | |
| ethanol | -* | + |
| lactose | V | + |
| salicin | - | + |

*-, 15% or less of strains positive; +, 85% or more positive; V, strain variation.

Table 6 summarizes data from several sources concerning Xanthomonas and P.paucimobilis. Gilardi /3/ has commented on the unreliability of host specificity for differentiation of species, and the history of Pasteurella multocida, Pseudomonas cepacia, and Brucella abortus strongly supports his view. Starr and his associates /11/ have shown that the taxa of Table 6 contain quite different pigments, and Moss /16/ has reported differences in their cellular fatty acids. However, since neither of these features is applicable to the clinical laboratory, we continue to believe that the distinction can usually be made on grounds of (1) source of specimen, (2) colonial morophology, and (3) acidification of ethanol and salicin.

Table 7. P.paucimobilis and IIk-2: Differential features

| | P.paucimobilis | IIk-2 |
|---|---|---|
| Colonies, 48 hr | | |
| pigment | yellow | none or pale yellow |
| size, mm | 1 | 2-3 |
| Acid from ethanol | +* | - |
| inulin | - | + |
| Assimilation of | | |
| glycerol | - | + |
| inulin | - | + |
| Urease | - | + |

*+, 90% or more of strains positive; -, 10% or less of strains positive.

By way of reiteration, we also submit that clinical bacteriologists should have no difficulty in distinguishing P.paucimobilis and IIk-2. Table 7 presents our observations in support of this conclusion, and Table 8 lists additional features of the IIk-2 group.

Table 8. IIk-2: Features of 22 strains

| | |
|---|---|
| **All positive** | |
| Acidification and assimilation of: | |
| L-arabinose, cellobiose, fructose, glucose, inulin, lactose, maltose, raffinose, salicin, corn and potato starch, sucrose, xylose | |
| Catalase | |
| Hydrolysis of: | |
| esculin, corn and potato starch, Tween 20 | |
| Oxidase | |
| Urease | |
| **All negative** | |
| Acidification of ethanol, inositol, mannitol, sorbitol | |
| Alkalinization of acetamide, citrate | |
| Arginine dihydrolase | |
| Assimilation of acetamide, D-arabinose, arginine betaine, ethanol, fucose, glycolate, inositol, lactate, maleate, mannitol, norleucine, putrescine, sorbitol, tryptophane, valine | |
| Decarboxylation of lysine, ornithine | |
| Growth, salmonella-shigella and Simmons' citrate agar | |
| Fluorescence | Motility |
| Indole | Nitrite reduction |
| Gluconate | Penicillin susceptibility |

Eighteen of the 22 strains of Table 8 appeared to be members of a single taxon as judged from the homogeneity of their phenotypic features and DNA hybridizations with representative strains (Table 9). However the 22 strains of IIk-2 were separable into two biovars by DNA hybridizations and by additional phenotypic features, as shown in Table 10. Note that the two biovars differ slightly in their moles% G+C, and they distinctly differ in their failure to hybridize. These two biovars are readily distinguished in tests for growth at 42C and reduction of nitrate to nitrite. Hence it appears that IIk-2 consists of more than one taxon. Other investigators /4,17,26/ have also reported strain variation within the IIk-2 group in respect to reduction of nitrate and growth at 42C.

Table 9. IIk-2: Moles% G+C and DNA hybridizations of selected strains

| Moles% G+C | Hybridizations, 65C | |
|---|---|---|
| | K-1205-$H^3$ | K-1226-$H^3$ |
| 41-45 (6 strains) | 74-100% (5 strains) | 77-100% (8 strains) |

Table 10. Biovars of IIk-2

| Biovar: | "A" | "B" |
|---|---|---|
| Number of strains: | 18 | 4 |
| Mucoid colonies | 0% | 75% |
| Growth at 42C | 0% | 100% |
| Nitrate → nitrite | 0% | 100% |
| Moles% G+C | 41-43 | 44-45 |
| DNA hybridizations | | |
| A with A | 74-100% | |
| A with B | | 0% |

Upon seeking a genus to which the IIk-2 group could be assigned, we considered Acinetobacter, Moraxella, and Flavobacterium since all, like IIk-2, are characterized by a G+C content of 35-45 moles%. Accordingly, we attempted, unsuccessfully, to effect hybridization of IIk-2 with one strain each of A.anitratus, M.phenylpyruvica, F.meningosepticum, and Flavobacterium group IIb. This negative hyrbidization was not surprising in the instance of Acinetobacter and Moraxella since their phenotypic features are quite different from those of IIk-2. Though we failed to detect hybridization of IIk-2 with the two strains of Flavobacterium, we do not consider this a strong argument against assigning IIk-2 to this genus since even within group IIa and within group IIb poor hybridizations have been reported /19,23/. On the contrary, we conclude that IIk-2 and flavobacteria do share a number of important features and that this group should be assigned to genus Flavobacterium. Common features are shown in Table 11. Both are human pathogens /2,20,25/. We believe the moles% of G+C in IIk-2, though slightly high, is compatible with the redefinition of genus Flavobacterium proposed by Holmes and Owen /6/. Our studies indicate that IIk-2 is an obligate aerobe, whereas some reports /25,28/ suggest that flavobacteria may be facultative. Price, however, found /22/ that all of the flavobacteria examined in her study were obligatively aerobic. To our knowledge, the yellow pigment

Table 11. Flavobacterium and IIk-2: Common features

| |
|---|
| Opportunistic pathogens for man |
| Similar moles% G+C |
| Gram negative rods |
| Obligatively aerobic |
| Yellow, nondiffusible pigment |
| "Nonfunctional" polar flagellum |
| Poor or no growth on MacConkey agar |
| Usually no growth at 42C |
| Catalase positive |
| Oxidase positive |
| Extracellular deoxyribonuclease |
| Menaquinone as repiratory pigment |
| Similar cellular fatty acid profiles |

produced by IIk-2 has not been compared chemically with the pigment(s) of flavobacteria. Until such time, we believe the pigmentation of IIk-2 should be considered am important feature shared with the flavobacteria. Though F.meningosepticum is nonmotile, Hugh's laboratory has reported /27/ that it appears to be monotrichously flagellated. Similarly, we have found that IIk-2, though nonmotile, bears a "floppy" polar flagellum (Figure 1). Holmes and Owen's /6/ revised definition of the genus Flavobacterium includes the statement, "Menaquinones are the sole respiratory quinones." We found /14/ that IIk-2 contains menaquinone and no, or at most a trace, of ubiquinone. As a final comparison, Moss' laboratory has reported /1/ that the cellular fatty acids of IIk-2 are similar to those of flavobacteria.

Table 12 is intended to summarize present taxonomy and nomenclature of the yellow-pigmented nonfermentative bacilli. Much progress has been made since the Roundtable of Elizabeth King in 1964. Many of her group designations have now been replaced by binomials. Her group I, now Pseudomonas maltophilia, was among the first of these /10/. The IIb group of flavobacteria remains without name(s). It is generally accepted that it is phenotypically related to F.meningosepticum, and Owen and Snell noted some DNA relatedness /19/. The London investigators have also reported /9/ that the CDC's "Moraxella-like" M-4f group and F.odoratum are the same taxon. An appropriate generic designation for the xanthomonads remains controversial. Studies by us and the investigators in London support the conclusion that IIk-2 should be assigned to genus Flavobacterium. Holmes, Owen, and Weaver /8/ have proposed the name F.multivorum for this group of yellow-pigmented bacteria. Holmes and his associates also recently assigned the name Pseudomonas paucimobilis to the IIk-1 group /7/. The P.stutzeri-complex (Vb) consists of P.stutzeri (Vb-1), P.mendocina (Vb-2), and Vb-3. Some clinical

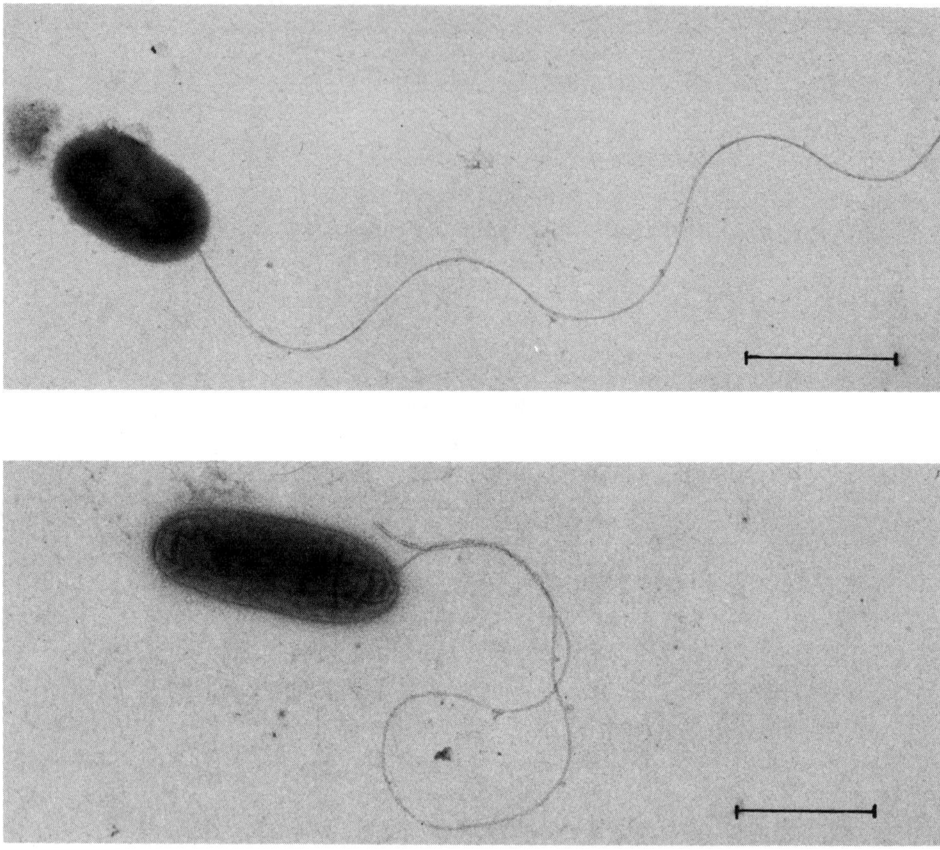

Figure 1. Electron micrographs of Flavobacterium group IIk-2, strain K-1205, showing (top) polar monotrichous flagellation and (bottom) that the flagella are often deformed and non-rigid. The flagella are about 20 nm in diameter. Micrographs courtesy of F.A. Eiserling, UCLA. The bar represents 1 μm.

bacteriologists consider strains of the Vb-3 group to be variants of P.stutzeri. Finally, the Ve or "'Chromobacterium' typhiflavum" group remains without official binomial(s) even though it has been extensively characterized /4,5,18,25/. The present binomial is illegitimate. Initially it included both facultatively and obligatively aerobic strains /20/. The facultative strains were later shown to be Enterobacteriaceae (Erwinia or Enterobacter agglomerans). Ve-1 and Ve-2 represent at least two taxa; they differ in flagellar morphology /5/, biochemical features /5/, cellular fatty acids /1/, and moles% G+C /5/.

| CDC's | % G+C | Flagella | Name |
|---|---|---|---|
| I | 67 |  | Pseudomonas maltophilia |
| EO-1 | 67-68 |  | Pseudomonas cepacia |
| IIa | 31-40 |  | Flavobacterium meningosepticum |
| IIb | 31-40 |  | Flavobacterium species |
| M-4f | 32-35 |  | Flavobacterium odoratum |
| IIk | | | Xanthomonas (? P. campestris) |
| IIk-2 | 41-43 |  | Flavobacterium multivorum |
| IIk-1 | 65-66 |  | Pseudomonas paucimobilis |
| Vb | 61-66 |  | Pseudomonas stutzeri complex |
| Ve-1 | 57 |  | "'Chromobacterium' typhiflavum" |
| Ve-2 | 67 |  | "'Chromobacterium' typhiflavum" |

Fig. 2. Yellow-pigmented gram negative rods: 1980.

## References


/ 1/ Dees, S.B., Moss, C.W., Weaver, R.E., Hollis D.: J. Clin. Microbiol. 10, 206-209 (1979).

/ 2/ Dhawan, V.K., Rajashekaraiah, K.R., Metzger, W.I., Rice T.W., Kallick, C.A.: J. Clin. Microbiol. 11, 492-495 (1980).

/ 3/ Gilardi, G.L. Health Lab. Sci. 12, 311-315 (1975).

/ 4/ Gilardi, G.L. p. 15-44 in: Gilardi, G.L. (ed.), Glucose nonfermenting gram-negative bacteria in clinical microbiology. West Palm Beach, Florida, USA (1978): CRC Press, Inc.

/ 5/ Gilardi, G.L., Hirschl, S., Mandel, M.: J. Clin. Microbiol. 1, 384-389 (1975).

/ 6/ Holmes, B., Owen, R.J.: Int. J. Syst. Bacteriol. 29, 416-426 (1979).

/ 7/ Holmes, B., Owen, R.J., Evans, A., Malnick, H., Willcox, W.R.: Int. J. Syst. Bacteriol. 27, 133-146 (1977).

/ 8/ Holmes, B., Owen, R.J., Weaver, R.E.: MS submitted to Int. J. Syst. Bacteriol.

/ 9/ Holmes, B., Snell, J.J.S., Lapage, S.P.: Int. J. Syst. Bacteriol. 27, 330-336 (1977).

/ 10/ Hugh, R., Ryschenkow, E.: J. Gen. Microbiol. 26, 123-132 (1961).

/ 11/ Jenkins, C.L., Andrewes, A.G., McQuade, T.J., Starr, M.P.: Current Microbiol. 3, 1-4 (1979).

/ 12/ King, E.O.: The identification of unusual pathogenic gram negative bacteria. Atlanta, Georgia, USA (1964): Communicable Disease Center.

/ 13/ Levine, M.G.: Nonfermentative bacilli: Phenotypic features, DNA-DNA hybridizations, and taxonomy of CDC's IIk Group. Los Angeles, CA, USA (1980): Doctoral Dissertation, UCLA

/ 14/ Levine, M.G., Pickett, M.J., Mandel, M.: Current Microbiology, in press.

/ 15/ Mandel, M.: J. Gen. Microbiol. 43, 273-292 (1966).

/ 16/ Moss, C.W.: p. 171-201 in: Gilardi, G.L. (ed.), op. cit.

/ 17/ Oberhofer, T.R.: J. Clin. Microbiol. 10, 800-804 (1979).

/ 18/ Oberhofer, T.R., Rowen, J.W., Cunningham, G.F.: J. Clin. Microbiol. 5, 208-220 (1977).

/ 19/ Owen, R.J., Snell, J.J.S.: J. Gen. Microbiol. 93, 89-102 (1976).

/ 20/ Pedersen, M.M., Marso, E., Pickett, M.J.: Am. J. Clin. Pathol. 54, 178-192 (1970).

/ 21/ Pickett, M.J., Pedersen, M.M.: Can. J. Microbiol. 16, 351-362 (1970).

/ 22/ Price, K.W.: A study of the taxonomy of flavobacteria isolated in clinical laboratories. Los Angeles, CA USA (1977): Doctoral Dissertation, UCLA.

/ 23/ Sottile, M.I., Baldwin, J.N., Weaver, R.E.: Appl. Microbiol. 26, 535-539 (1973).

/ 24/ Stanier, R.Y., Palleroni, N.J., Doudoroff, M.: J. Gen. Microbiol. 43, 159-271 (1966).

/ 25/ Tatum, H.W., Ewing, W.H., Weaver, R.E.: p. 270-294 in: Lennette, E.H., Spaulding, E.H., Truant, J.P. (eds.), Manual of clinical microbiology. Washington, D.C., USA (1974): American Soc. Microbiol.

/ 26/ Weaver, R.E., Tatum, H.W., Hollis, D.G.: The identification of unusual pathogenic gram negative bacteria. Atlanta, Georgia, USA (1972): Center for Disease Control.

/ 27/ Webster, J.A., Hugh, R.: Int. J. Syst. Bacteriol. 29, 333-338 (1979).

/ 28/ Weeks, O.B.: p.357-364 in: Buchanan, R.E., Gibbons, N.E. (eds.), Bergey's manual of determinative bacteriology, 8th ed. Baltimore, Maryland, USA (1974): Williams & Wilkins Co.

# STUDIES OF CLINICAL ISOLATES OF FLAVOBACTERIA

K. W. Price*, M. J. Pickett
Department of Microbiology
University of California, Los Angeles, CA 90024, USA

## Summary

Computer analysis was used in a numerical taxonomy study of flavobacteria from clinical, environmental, and marine sources. The 138 strains examined included representatives of the Center for Disease Control (CDC) groups IIa, IIb, and IIf. None of the marine strains was similar to those of clinical origin. The largest group, IIb of the CDC and Group O of McMeekin, appeared to merit recognition as a named species. Group IIa (*Flavobacterium meningosepticum*) was similar to group IIb, and was antigenically related to group IIf.

## Introduction

In 1970 Pickett and his associates published a series of papers on nonfermentative bacilli isolated in clinical laboratories /12-15/. They assigned these nonfermenters to 8 genera, of which genus *Flavobacterium* presented the greatest taxonomic problem. They found three groups which could be placed in this genus. Their first group was *Flavobacterium meningosepticum* (CDC Group IIa). This species was first recognized by King /8/ as the causative agent of epidemic meningitis among newborn infants. This species is rarely isolated. The only strain of this species included in their study was ATCC 13253. Their second group of flavobacteria was unnamed and is known as CDC Group IIb. It is an occasional isolate from clinical specimens. Their third group, also unnamed, is known as CDC Group IIf.

In other areas of microbiology flavobacteria have presented similar problems. Hayes /3/ studied marine flavobacteria. McMeekin and his associates studied flavobacteria isolated from dairies and food processing plants /10/. Hayes and McMeekin found heterogeneity among their strains. Their groups did not correspond to any named species.

*Present address: Cedars-Sinai Medical Center, Los Angeles, CA 90048.

The specific goals of our research were:

1) to find differences between F.meningosepticum and the other flavobacteria isolated from clinical specimens;

2) to study CDC Group IIb, looking for subgroups within that group and determining if it should be given species rank;

3) to study CDC Group IIf, looking at its similarity to F.meningosepticum and determining if it should be given species rank; and

4) to compare the groups of clinical flavobacteria with the groups found by Hayes and by McMeekin.

## Materials and Methods

### Bacterial strains.

We examined 115 flavobacteria-like clinical isolates, listed in Table 1. We also included 14 clinical isolates which were not flavobacteria. They are listed in Table 2. Table 3 lists the reference strains of marine and food flavobacteria which were included in the study.

Table 1. Sources of flavobacterial cultures

| Donor | No. of strains | Locale |
|---|---|---|
| E. Marso | 35 | Los Angeles, CA. |
| D. Lambe | 30 | Atlanta, GA. |
| R. Weaver | 19 | Atlanta, GA. |
| G. Gilardi | 15 | New York, NY |
| G. Nygaard | 5 | Berkeley, CA |
| A. von Graevenitz | 4 | New Haven, CT |
| Others | 7 | Various |

Table 2. Control cultures

| | No. of strains |
|---|---|
| Nonfermentative, pigmented | |
| Pseudomonas aeruginosa | 1 |
| Pseudomonas putrefaciens | 3 |
| CDC Group IIk | 3 |
| CDC Group Ve | 3 |
| Nonfermentative, nonpigmented | |
| Acinetobacter anitratus | 2 |
| Fermentative | |
| Escherichia coli | 1 |
| Klebsiella pneumoniae | 1 |

Table 3. Nonclinical flavobacteria included in the study

| | |
|---|---|
| Hayes Group II | McMeekin Group O |
| NCMB 251 | Strain 25 |
| NCMB 264 | Strain 42 |
| NCMB 289 | |
| | McMeeking Group Y |
| Hayes Group III | Strain 13 |
| NCMB 249 | |
| | McMeekin Group S |
| Hayes Group IV | Strain 60 |
| NCMB 275 | |

Morphology and biochemical, growth, and antimicrobial sensitivity tests.

Over 150 tests were performed on each culture. Complete details are given elsewhere /16/. The results of 128 tests were used for the calculation of percent similarity (%S). Table 4 summarizes these characters.

Table 4. Characters used for data analysis

| | | Number of characters |
|---|---|---|
| Gross morphology | | 26 |
| Appearance of broth | | 7 |
| Microscopic morphology | | 17 |
| Growth tests | | 12 |
| Antimicrobial sensitivities | | 16 |
| Buffered single substrate tests | | 25 |
| sugars | 19 | |
| alcohol | 1 | |
| amides | 2 | |
| salts | 3 | |
| Enzymes | | 9 |
| Miscellaneous | | 16 |
| Total: | | 128 |

Data analysis.

The calculation of %S was made using the formula:

$$\%S = 100 \times \frac{NS}{NS + ND}$$

where NS or number similar is the number of characters for which the two strains have the same result and ND or number different is the number of characters for which the two strains have different results. The %S data were used for grouping the strains. Two grouping methods were used. The first was single linkage analysis. The second was a program which we wrote and called complete linkage analysis. The logic for complete linkage analysis was derived from E.W. Fager /2/ who used it to study the flexibacteria:

> "The programme first formed the largest group within which all possible pairs of strains had affinity indices at or above a preset value. The members of the first group were then removed from further consideration and the largest possible group was formed among the remaining strains."

Complete linkage analysis uses two criteria for group formation. The first criterion is homogeneity. All affinity indices, or in this case %S, must be equal to or greater than a preset value, which we have called minimum percent similarity (MPS). The second criterion is size. The program selects the largest possible group which meets the requirement for homogeneity.

## Results

Common features for a majority of strains are shown in Table 5. They are gram negative rods. They fail to acidify the butt of Kligler's iron agar and therefore, for practical purposes, are considered nonfermentative. They are pigmented, oxidase positive and catalase positive. All but two of our strains are obligately aerobic, that is they fail to grow on plates in a "Gas-pak" jar. Only two strains are motile in hanging drop preparations; these two are atypical in many other features.

Table 5. Common features for most strains

| Feature | Positive # | Positive % |
|---|---|---|
| Gram negative rods | 115 | 100 |
| Nonfermentative | 115 | 100 |
| Pigmented | 115 | 100 |
| Oxidase | 115 | 100 |
| Catalase | 115 | 100 |
| Motility | 2 | 2 |
| Anaerobic growth | 2 | 2 |

Single linkage analysis.

The matrix in Figure 1 and the dendrogram in Figure 2 illustrate the percent similarity data. Only 132 strains could be included in the matrix because the computer can print only 132 characters per line. The strains omitted are *E. coli*, *K. pneumoniae*, *A. anitratus* (2 strains), *P. aeruginosa*, and one of the motile clinical isolates. The dendrogram includes all 138 strains. The strains are listed in the order determined by the single linkage analysis program. Limits which we chose for the three major groups are indicated. The largest group is B.

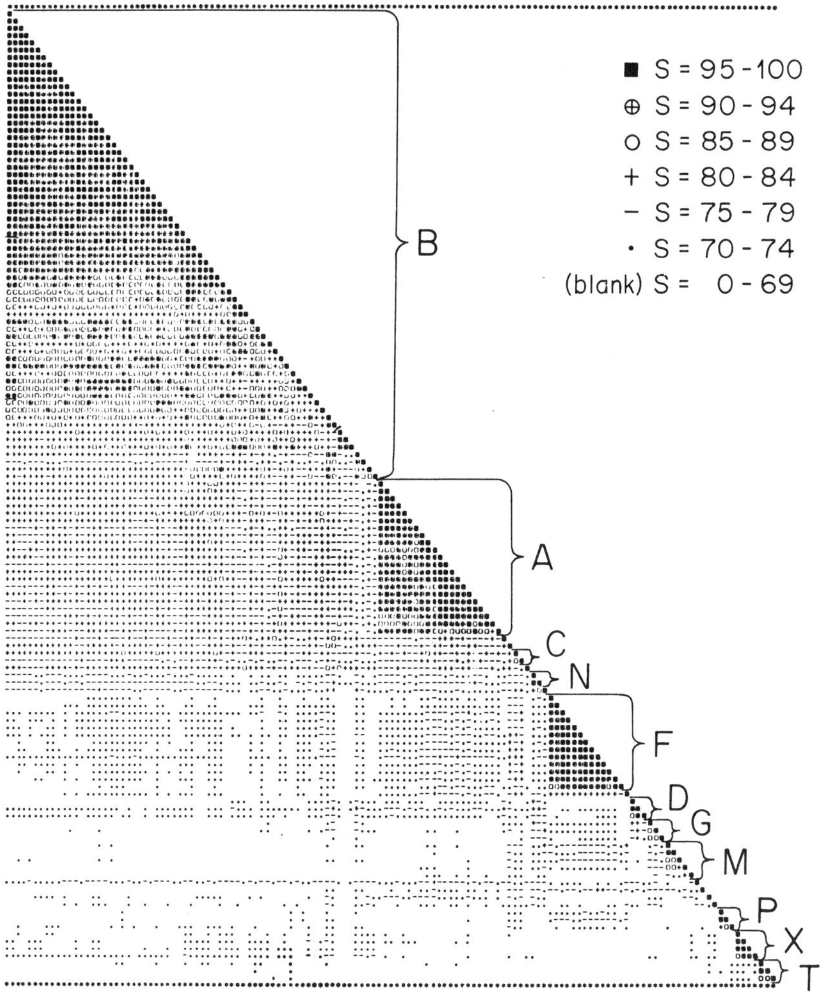


FIGURE 1. MATRIX OF % S.

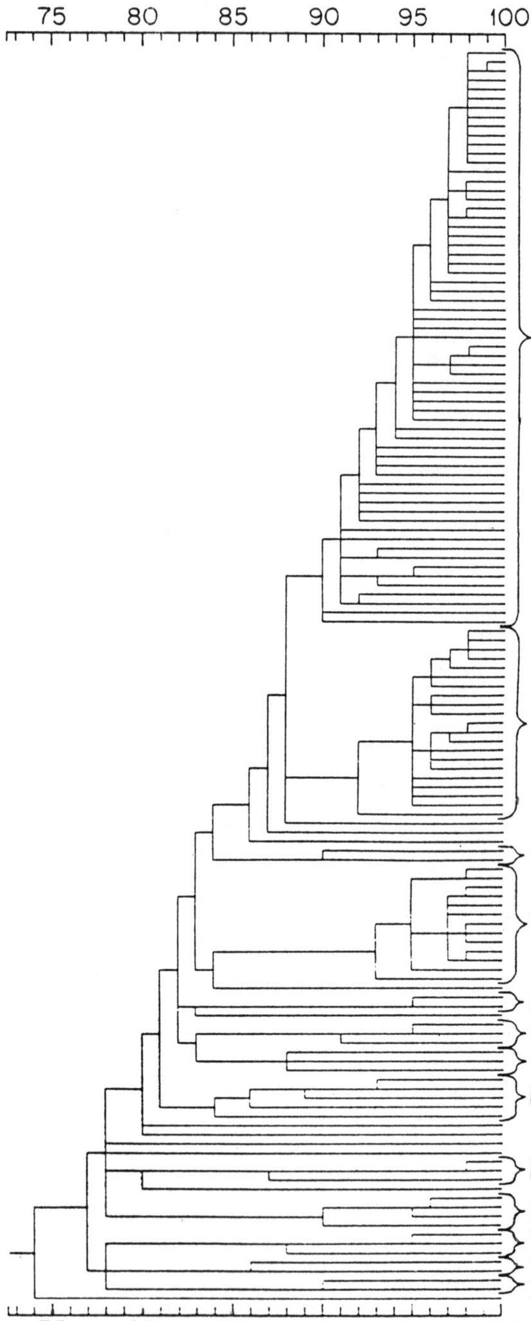


FIGURE 2. RESULTS OF SINGLE LINKAGE ANALYSIS.

It includes 63 clinical strains and one of McMeekin's strains. It is equivalent to CDC Group IIb. Within the group, Figure 1 shows some high %S and some lower %S values. These lower values indicate heterogeneity in the group, but no subgroups can be found in this matrix. Our Group A is CDC Group IIa, F. meningosepticum. The number of strains included here does not reflect the actual frequency of isolation. This species is rarely isolated. The matrix illustrates that this group is distinct and %S within the group are high. Our Group F is CDC Group IIf; it is also distinct and has high %S within the group.

In addition to the three major groups, four minor groups can be formed. They are Group C, two strains placed between Groups A and F; Group N, two strains below Group F; and Groups D and G, three strains each, which are first clustered together, then joined to the main cluster below Group N. Seven clinical isolates are left as single strains. One of these single strains, which is placed immediately after Group F, is probably Flavobacterium odoratum.

The marine flavobacteria form Group M which is linked to the main cluster below Groups D and G. The four strains of flavobacteria isolated from food are placed as follows: food strain 25, McMeekin Group O, is placed in Group B. It is the second to the last strain included in the group. Food strain 42, also McMeekin Group O, is the first strain after Group A. The %S data indicate that this strain is more similar to Group B than to Group A. Food strain 60, McMeekin Group S, and food strain 13, McMeeking Group Y, are among 4 single strains which are joined to the main cluster below Group M. By our methods food strain 13 is a gram positive organism. All of the nonflavobacterial clinical isolates were linked to the main cluster below the marine and food strains. They are shown in the dendrogram as Group P, X, T, Y and Z and the last single strain, which is P.aeruginosa.

Description of the groups

Group A, Flavobacterium meningosepticum, grows moderately well. The pigment is a pale cream color. Table 6 presents salient features of this species. It is homogeneous in respect to its acidifcation of carbohydrates. The key positive reactions are hydrolysis of gelatin and production of indol. Indol production is weak, as originally noted by King /8/. Table 7 presents antimicrobial sensitivities of F.meningosepticum, which would be more appropriately called antimicrobial resistances. This species is completely resistant to 12 of the 16 antimicrobials which we tested.

Table 6. Flavobacterium meningosepticum
Features of 21 strains

| | | | |
|---|---|---|---|
| Acidification of | | Acidification of | |
| D-arabinose | 100* | salicin | 0 |
| L-arabinose | 0 | sorbitol | 0 |
| cellobiose | 100 | sucrose | 0 |
| fructose | 100 | trehalose | 100 |
| galactose | 0 | Alpha hemolysis | 0 |
| glucose | 100 | Beta hemolysis | 0 |
| inositol | 0 | Amylase | 0 |
| lactose | 100 | DNase | 100 |
| maltose | 100 | Gelatin | 100 |
| mannitol | 100 | Gluconate | 0 |
| mannose | 100 | Growth at 42C | 0 |
| melibiose | 100 | Growth in 6% NaCl | 0 |
| raffinose | 0 | Indol | 100 |
| rhamnose | 0 | Nitrate reduction | 0 |
| | | Phenylalanine (PPA) | 0 |

*% of strains positive.

Table 7. Flavobacterium meningosepticum
Antimicrobial sensitivities of 21 strains

| | | | |
|---|---|---|---|
| Chloramphenicol | 19% | Bacitracin, cephalothin, colymycin, furadantin, gentamycin, kanamycin, neomycin, penicillin, polymyxin B, streptomycin, sulfadiazine, tetracycline | 0% |
| Erythromycin | 95% | | |
| Nalidixic acid | 33% | | |
| Vancomycin | 100% | | |

Group B, in contrast to Group A, is heterogeneous. It grows well. The pigment is usually a striking golden to orange color, but a few strains produce only cream colored pigment. A unique feature of this group is production of a slime layer, as described by McMeekin et al. /10/. This slime layer was produced by 87% of our Group B strains; no other strain in our study showed this phenomenon. Additional features of Group B are presented in Tables 8 and 9. Note that only 95% of the strains were indol positive. The negative indol reactions were confirmed by repeated tests. The carbohydrate reactions were similar to those for F.meningosepticum, but there was marked strain variation. Sensitivity to antimicrobials was also similar to that for F.meningosepticum, but there was more strain variation.

Table 8. Features of Flavobacterium Group B (63 strains)

| | | | |
|---|---|---|---|
| Acidification of | | Acidification of | |
| D-arabinose | 0* | salicin | 6 |
| L-arabinose | 25 | sucrose | 14 |
| cellobiose | 88 | trehalose | 100 |
| fructose | 98 | xylose | 37 |
| galactose | 0 | Alpha hemolysis | 51 |
| glucose | 100 | Beta hemolysis | 4 |
| lactose | 0 | Amylase | 98 |
| maltose | 100 | DNase | 31 |
| mannitol | 7 | Gelatin | 100 |
| mannose | 100 | Gluconate | 9 |
| melibiose | 90 | Indol | 95 |
| raffinose | 0 | Nitrate reduction | 30 |
| rhamnose | 1 | Slime layer | 87 |

*% positive.

Table 9. Flavobacterium Group B
Antimicrobial sensitivities of 63 strains

| | | | |
|---|---|---|---|
| Erythromycin | 41* | Streptomycin | 13 |
| Nalidixic acid | 97 | Tetracycline | 5 |
| Vancomycin | 98 | Colymycin | 0 |
| Cephalothin | 5 | Kanamycin | 0 |
| Chloramphenicol | 13 | Penicillin | 0 |
| Neomycin | 17 | Polymyxin B | 0 |

*% sensitive.

Because of the heterogeneity within Group B we tried to find subgroups within the overall group. As mentioned earlier, no subgroups were found by single linkage analysis. Another approach was analysis of test results. We believe that division into subgroups should be based on two or more distinguishing tests. We could not find even two tests to use as a basis for dividing strains into subgroups. For example, when strains were divided into DNase positive and DNase negative subgroups, these subgroups were still variable for beta hemolysis, nitrate reduction, phenylalanine deaminase, and acidification of L-arabinose, xylose, and ethanol.

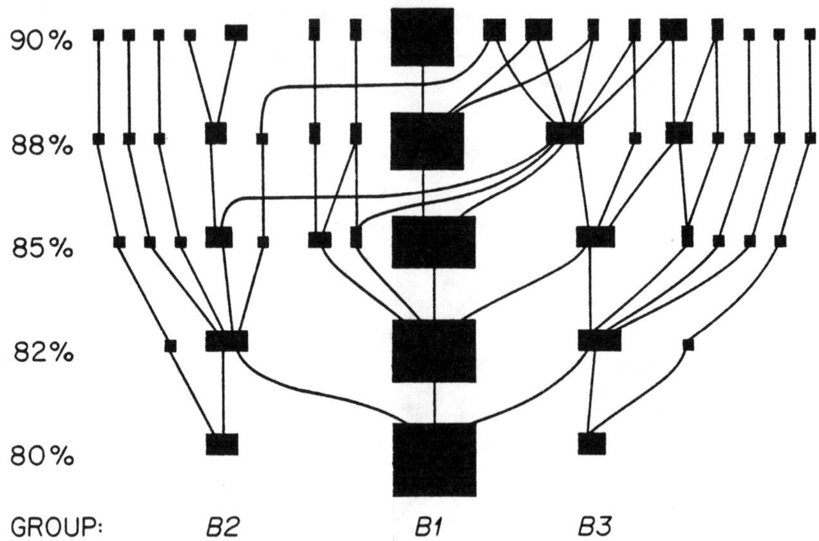


FIGURE 3. RESULTS OF CLA FOR 67 STRAINS.

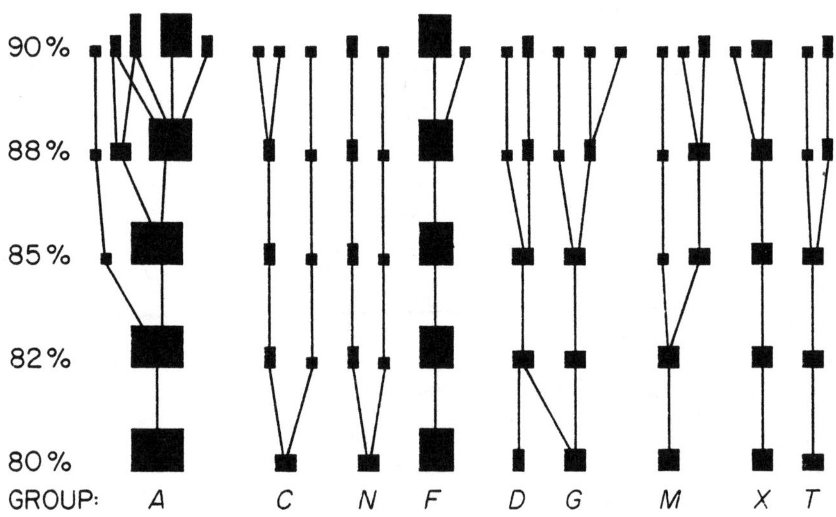


FIGURE 4. RESULTS OF CLA FOR 57 STRAINS.

Our third approach to finding subgroups within Group B was the complete linkage analysis program (CLA). The results of CLA are presented in Figures 3 and 4. The program was run using 5 different values for minimum percent similarity (MPS). The MPS is shown at the left in Figures 3 and 4. The groups formed at a particular MPS are represented by blocks in a horizontal row to the right of the MPS value. The size of each block is proportional to the number of strains in the group. The vertical lines connecting the blocks have been added to show how strains change groups as the MPS is changed. When the MPS is lowered groups may merge and individual strains may be added to a group, removed from a group or be placed into a different group. The results for 57 strains not in Group B will be discussed first as an introduction to this method of group formation.

When the MPS is 85%, these groups correspond to the groups formed from the single linkage analysis program, except for one strain of Group A and the five strains of marine flavobacteria (Group M). When the MPS is 82%, these groups correspond to the single linkage analysis groups. When the MPS is 80%, Groups D and G begin to merge and single strains are added to Groups C and N.

Figure 4 presents the results of CLA for 67 strains of Group B. These 67 strains include the 64 placed in Group B by single linkage analysis plus two more clinical isolates and a second strain of flavobacteria from food. From the results of CLA we considered 3 possible subgroups of Group B. Figure 4 illustrates that these subgroups are not distinct. They merge and divide with each change in the MPS. Intermediate strains exist which are in subgroup B3 at one MPS and in subgroup B1 at a lower MPS. There are also equivocal strains whose %S values would permit them to be placed in 2 different subgroups without changing the MPS. We examined test profiles of these subgroups and found only one or two tests which distinguish them.

Group F strains grow more slowly than do strains of F.meningosepticum or Group B. Colonies are less than 1 mm in diameter after one day, but grow to 5-10 mm after several days. They are mucoid and have a pale yellowish color. Table 10 summarizes their features. All grew at 42C and produced indol, phenylalanine deaminase, gelatinase and DNase. They are nonsaccharolytic; the negative reaction to glucose shown in the table is representative of all carbohydrates and alcohols. They are more sensitive to antimicrobials than F.meningosepticum and Group B, notably, to cephalothin and penicillin.

Table 10. Flavobacterium Group F: Salient features of 13 strains

| | | | |
|---|---|---|---|
| Growth at 42C | 100* | Sensitive to | |
| Alpha and beta hemolysis | 0 | bacitracin, chloramphenicol | |
| Gluconate | 0 | cephalothin, erythromycin | |
| Nitrate reduction | 0 | penicillin, vancomycin | 100 |
| Indol | 100 | nalidixic acid | 92 |
| Phenylalanine (PPA) | 100 | neomycin | 7 |
| Gelatin | 100 | polymyxin B | 15 |
| Amylase | 0 | gentamicin, kanamycin, | |
| DNase | 100 | streptomycin | 0 |
| Glucose | 0 | | |

*% positive.

Table 11. Flavobacteria: Comparison of groups

| Group: | A | B | F |
|---|---|---|---|
| Number of strains: | 21 | 63 | 13 |
| Pigment | Cream | Orange | Cream |
| | % Positive | | |
| Slime layer | 0 | 87 | 0 |
| Growth at 42C | 0 | 4 | 100 |
| Phenylalanine | 0 | 20 | 100 |
| Amylase | 0 | 98 | 0 |
| Lactose | 100 | 0 | 0 |
| d-Arabinose | 100 | 0 | 0 |
| Mannitol | 100 | 7 | 0 |
| Glucose | 100 | 100 | 0 |
| Cephalothin | 0 | 5 | 100 |
| Penicillin | 0 | 0 | 100 |
| Tetracyline | 0 | 5 | 92 |

Comparison of groups

Table 11 summarizes the differences among the three major groups of clinical flavobacteria. As noted by Brody et al. /1/, F.meningosepticum and CDC Group IIb are very similar. The tests which we found to be the most reliable and useful for differentiating them are pigmentation, slime layer, amylase, lactose and mannitol. CDC Group IIf is distinguished from the other two groups by its negative reactions to all sugars and by its sensitivity to penicillin, cephalothin and tetracycline.

Serology

There was cross agglutination among the three clinical groups. For example, one anti-Group A serum agglutinated 5 of 6 Group A strains, 1 of 11 Group B strains, and 1 of 2 Group F strains. Another anti-Group A serum agglutinated 3 of 6 Group A strains, 10 of 11 Group B strains and neither of 2 Group F strains. Similarly anti-Group B sera agglutinated some strains of each of the three groups. Finally, anti-Group F sera agglutinated both strains of Group F, and some strains of Groups A and B.

## Discussion and Conclusions

Group A.

Our results are in agreement with King's description of Flavobacterium meningosepticum /8/. This species is phenotypically homogeneous and distinct.

Group B.

This group is phenotypically heterogeneous. Should it be divided into subgroups? It is our belief that any taxonomic division is warranted only if it reflects a naturally occuring division of strains into separate clusters. The majority of strains should fall into one or another of the taxonomic units to be created. For Group B, clusters exist but they are not distinct. There are numerous intermediate strains which have high similarities to members of different clusters. This view of Group B is supported by the three different methods of data analysis which we have used. Heterogeneity within Group B is recognized, but any division into subgroups would be artificial and arbitrary.

We recommend the recognition of CDC Group IIb as a single species within genus Flavobacterium, and suggest F.aureum as an apt name for this species.

Group F.

This group, like A, is phenotypically homogeneous and distinct. We recommend its recognition as a species within genus Flavobacterium. Because it is most frequently isolated from the genitourinary tract, we propose the name Flavobacterium genitale. We designate NCTC 10799 (CDC's B 912; our K-823) as the type species.

Comparison of flavobacteria from marine, food, and clinical sources.

Results from three studies of flavobacteria are shown in Table 12. Each horizontal row in the table represents a single group. Marine Group A is now recognized as weakly gram positive (Shewan, personal communication) and is therefore excluded from genus Flavobacterium. McMeekin's results /11/ indicate that his food Group Y is the same as marine Group A. Our results show that McMeekin's Group O is the same as CDC's Group IIb.

In 1969 the genus Flavobacterium was described by Weeks /17/ as "...more a historic concept than a taxonomic reality." From our studies and those of Hayes and McMeekin, 7 major groups of flavobacteria can be recognized. Since our studies were completed, the work of Holmes and his associates and of Levine have led to recognition of additional specific groups of flavobacteria /4-7,9/. The genus Flavobacterium is becoming a taxonomic reality.

Table 12. Comparison of flavobacteria from different sources

| Marine Groups | Food Groups | Clinical Groups |
|---|---|---|
| A* | Y | - |
| T | - | - |
| G | - | - |
| - | O | IIb |
| - | S | - |
| - | A | - |
| - | - | IIa |
| - | - | IIf |

*A, marine Group A; -, not found.

## References

/1/ Brody, J.A., Moore, H., King, E.O.: Am. J. Dis. Child. 96, 1-5 (1958).

/2/ Fager, E.W.: J. Gen. Microbiol. 58, 179-187 (1969).

/3/ Hayes, P.R.: J. Gen. Microbiol. 30, 1-19 (1963).

/4/ Holmes, B., Owen, R.J., Weaver, R.E.: Int. J. Syst. Bacteriol. Submitted for publication.

/5/ Holmes, B., Snell, J.J.S., Lapage, S.P.: Int. J. Syst. Bacteriol. 27, 330-336 (1977).

/6/ Holmes, B., Snell, J.J.S., Lapage, S.P.: Int. J. Syst. Bacteriol. 28, 201-208 (1978).

/7/ Holmes, B., Snell, J.J.S., Lapage, S.P.: J. Clin. Pathol. 32, 73-77 (1979).

/8/ King, E.O.: Am. J. Clin. Pathol. 3, 241-247 (1959).

/9/ Levine, M.G., Pickett, M.J., Mandel, M.: Current Microbiol. In press.

/10/ McMeekin, T.A., Patterson, J.T., Murray, J.G.: J. Appl. Bacteriol. 34, 699-716 (1971).

/11/ McMeekin, T.A., Steward, D.B., Murray, J.G.: J. Appl. Bacteriol. 35, 129-137 (1972).

/12/ Pedersen, M.M., Marso, E., Pickett, M.J.: Am. J. Clin. Pathol. 54, 178-192 (1970).

/13/ Pickett, M.J., Manclark, C.R.: Am. J. Clin. Pathol. 54, 155-163 (1970).

/14/ Pickett, M.J., Pedersen, M.M.: Can. J. Microbiol. 16, 351-362 (1970).

/15/ Pickett, M.J., Pedersen, M.M.: Can. J. Microbiol. 16, 401-409 (1970).

/16/ Price, K.W.: A study of the taxonomy of flavobacteria isolated in clinical laboratories. Los Angeles, CA, USA (1977): Doctoral Dissertation, UCLA.

/17/ Weeks, O.B.: J. Appl. Bacteriol. 32, 13-18 (1969).

# CLASSIFICATION OF GROUP II K - 2 AND RELATED BACTERIA

E. Yabuuchi[1], I. Yano[2], T. Kaneko[3] and A. Ohyama[1]

[1]Kansai Medical University, Department of Microbiology, Moriguchi, Osaka 570, [2]Osaka City University, School of Medicine, Department of Bacteriology, Abeno-Ku Osaka 545, and [3]Institute of Physical and Chemical Research, Wako-Shi, Saitama 351, Japan

## Summary

Eleven strains of group II k - 2 and related bacteria were studied for their phenotypic characters, guanine plus cytosine content of their deoxyribonucleic acid and cellular lipid composition, in comparison with those of 41 strains of 5 *Flavobacterium* species and the type strain of *Cytophaga johnsonae*. The 11 strains were nonmotile, nonfermentative, Gram-negative, oxidase-positive rods. The GC content of DNA was 38.9 ± 0.9 mole%. Numerical analysis of 92 phenotypic characters indicated that the cluster of the 11 strains was rather close to *C. johnsonae* but clearly differed from 3 other major clusters: *Flavobacterium meningosepticum*, *Flavobacterium* II b and *Flavobacterium odoratum* / *breve* group. Major fatty acids of total extractable lipids of II k - 2 organisms were 13-methyltetradecanoic acid and 2-hydoxy 13-methyltetradecanoic acid. The cellular lipids of 11 strains contained sphingophospholipids which was so far not found in *Flavobacterium* sp. and *C. johnsonae* strains. The most abundant molecular species of ceramides of the sphingophospholipids was determined as being composed of branched chain C17 sphinganine and 2-hydroxy 13-methyltetradecanoic acid, exclusively. Based on the results of numerical analysis and presence of this unique sphingophospholipid, we are going to propose the new name, *Sphingobacterium* gen. nov. and *Sphingobacterium versatilis* sp. nov. for the group of strains, including those previously called group II k - 2, according to the International Code of Nomenclature of Bacteria. The strain KM 2138 will be designated as the type strain for *Sphingobacterium versatilis* sp. nov, type species of the genus. Strains of *Flavobacterium* sp. group II b are regarded as conforming to a separate species.

## Introduction

Sphingolipids are known to be important components of animal cell membranes. In bacteria, however, species of certain limited genera have been reported to have this kind of compound /7, 11, 12, 13, 17, 20, 23, 24/. In 1978, the cellular lipids of Flavobacterium devorans ATCC 10829 were reported to contain a new sphingoglycolipid, ceramide glucuronic acid /26/. Presence of this unique compound in the cellular lipids of the type and reference strains of Pseudomonas paucimobilis lead to the reidentification of ATCC 10829 as a strain of P. paucimobilis /25/, although the culture is still documented as the nomenclatural type of the species in the Approved Lists of Bacterial Names /19/.

Group II k organisms were divided into 2 biotypes, 1 and 2 /22/, and the former was included in the species P. paucimobilis /10/. Strains very much like the latter, II k - 2, have recently frequently been isolated from clinical materials in Japan and has caused purulent meningitis in a premature infant /14/. The 11 strains, including two II k - 2 strains, were uniformly nonmotile, assimilated many kinds of carbohydrates and differed from group species of reference strains used in their phenotypic characters. Furthermore, the cellular lipids of the 11 strains contained branched chain fatty acids in high concentration and sphingophospholipids /27/. Because of the clinical significance and taxonomic interest, 11 strains were characterized in comparison with 41 strains of 5 Flavobacterium species and the type strain for C. johnsonae, and the names Sphingobacterium gen. nov. and Sphingobacterium versatilis sp. nov. will be proposed for the 11 strains of II k - 2 organism.

## Materials and Methods

The 11 strains of S. versatilis are listed in Table 1. Two CDC (Center for Disease Control) strains of II k - 2 were included. Others were received as strains of Flavobacterium species, Flavobacterium meningosepticum, or Xanthomonas campestris. Strain KM (Kansai Medical University) 2087, received as X. campestris, was isolated from turnip, and all other 10 strains were of clinical origin. Thirteen strains of Flavobacterium II b, 18 F. meningosepticum, 5 Flavobacterium odoratum, 4 Flavobacterium breve, Flavobacterium saccharolytica ATCC 19248 (= KM 1933), and C. johnsonae ATCC 17061 (= KM 2500) were used as references, because they are all oxidase-positive nonfermenters of low GC content. Strain ATCC 19248, isolated from spinal fluid of a child with meningitis, was named and described by Flamm /5/ as Moraxella saccharolytica. Lautrop labeled the culture as Flavobacterium saccharolytica and deposited it with the American Type Culture Collection.

Casitone (Difco, 1.0%)-yeast extract (Difco, 0.3%)(CY) broth, containing 0.5% sodium chloride, was routinely used for growing these organisms. Cultures were incubated

Table 1. The 11 strains of Sphingobacterium versatilis sp. nov.

| KM | Received as | Received from | Isolated from |
|---|---|---|---|
| 926 | Group II K - 2, CDC B3159 | R. Hugh | Blood |
| 1203 | Flavobacterium meningosepticum | S. Mizuta | Spinal fluid |
| 2055 | Flavobacterium sp. No.22 | K. Karino | Clinical specimen |
| 2087 | Xanthomonas campestris AJ2779 | K. Komagata | Turnip |
| 2120 | Flavobacterium meningosepticum | W. Minamide | Synovial fluid |
| 2134 | Flavobacterium sp. No.2 | W. Minamide | Vaginal secretion |
| 2138 | Flavobacterium sp. No.6 | W. Minamide | Vaginal secretion |
| 2148 | Flavobacterium sp. No.9 | W. Minamide | Vaginal secretion |
| 2274 | Flavobacterium sp. No.7 | N. Kosakai | Urine |
| 2289 | Flavobacterium sp. No.28 | N. Kosakai | Sputum |
| 2532 | Group II K - 2, CDC A8895 | R. Hugh | Cavity fluid |

AJ Ajinomoto Central Laboratory, Kawasaki, Japan
CDC Center for Disease Control, Georgia, U.S.A
KM Kansai Medical University, Osaka, Japan

at 30C unless otherwise stated. Sliding motility was observed on a plate of CY agar (0.5%). Each organism was inoculated at the center of the plate which was maintained at room temperature. Deoxyribonuclease activity was tested by means of DNA agar slant containing 0.01% toluidine blue (Poa-media, Eiken, Tokyo). Anaerobic respiration by nitrate, nitrite, or fumarate was determined by testing anaerobic growth in CY broth supplemented with 0.1% potassium nitrate, sodium nitrite, or 0.08% sodium fumarate /1/. One percent tryptone broth supplemented with 0.05% L-tryptophane was used for indole production. Skim milk powder 5g, yeast extract 0.5g, sodium acetate 0.2g, and agar 15g were used to make one liter of skim acetate agar /2/. Since the deep yellow growth of C. johnsonae and Flavobacterium II b strains on the surface of semisolid VP medium became deep red when Voges-Proskauer test reagents were added, alkaline color change was tested by adding 20% potassium hydroxide solution to the colonies of 53 test strains grown on blood agar base (Difco 0045) medium. Other methods used for characterization of 53 strains have been described previously /25/.

Susceptibility of 53 strains against the following 14 antimicrobial agents was tested by agar diffusion technique; penicillin G, ampicillin, carbenicillin, cephazolin, gentamicin, kanamycin, tobramycin, tetracycline, minocycline, erythromycin, chloramphenicol, colistin, and nalidixic acid. Blood agar base and Showa discs (Showa Yakuhin Kako, Tokyo) were used.

For numerical analysis of phenotypic characters, data on such characters that all 53 strains showed uniformly positive or uniformly negative reactions were excluded, and the simple matching coefficients and average linkage method were used. The results of susceptibility testing were also excluded.

For lipid analysis, organisms were grown on peptone (Difco,1.0%)-yeast extract (0.5%)-glucose (1.0%)-agar (1.5%). After 20h incubation at 30C, the cells were harvested and the lipids were extracted twice with 20 vol. of chloroform:methanol (2:1, by vol.) and washed by the method of Folch et al./6/. The lipids were hydrolyzed with 0.5N KOH in methanol-chloroform (1:1, by vol.) for 2 to 6h at room temperature with occasional shaking. After the reaction mixture was neutralized, the resultant fatty acids and alkaline stable lipids were extracted with chloroform-methanol (2:1, by vol.). Acid methanolysis was carried out with methanol-12N HCl (5:1, by vol.) at 100C for 3h /21/. After cooling, an equal volume of water was added and the fatty acid methyl esters were extracted twice with n-hexane. The aqueous phase was then made to pH 12 with KOH pellets and the long-chain bases were extracted twice with n-hexane-diethyl ether (1:1, by vol.). The ceramides were prepared by HF hydrolysis from alkaline stable phospholipids according, essentially, to the method of Shaw and Stead /18/. The lipids were separated by thin-layer chromatography and the phospholipids were visualized by spraying with Dittmers' reagent and the long-chain bases with ninhydrin reagent. Gas-liquid chromatography of fatty acid methyl esters was carried out on a Hitachi 063 apparatus with a column of 15% diethyleneglycolsuccinate polyester or 5% SE-30. Trimethylsilylether derivatives of the long-chain bases (or intact ceramides) were prepared with trimethylchlorosilane and hexamethyldisilazane in pyridine /8/, and separated gas-chromatographically on a 5% SE-30 column at 210C. Gas chromatographic and mass spectrometric analysis were carried out with the Hitachi AMU 6MG apparatus.

## Results and Discussion

The 11 strains of S. versatilis were Gram-negative, asporogenous rods. Motility was not demonstrated either on a Gard plate or under the microscope. Flagella were not visualized by Leifson flagella stain. Capsule and appendages were not observed. They produced pale yellow colonies on heart infusion agar plates at either 25C or 37C. They were cytochrome oxidase, catalase and phosphatase positive, and hydrolysed esculin. Oxidative acidity was produced in OF basal medium from pentoses, hexoses, disaccharides, trisaccharides and polysaccharide. They had sphingophospholipids as their cellular lipid component, and were susceptible to clindamycin and minocycline. All 11 strains failed to produce indole, a clear zone around the growth on skim milk acetate agar and acid from sorbitol, dulcitol and isositol. Their growth was not inhibited with 5 μg colistin, 30 μg tobramycin, and 30 μg gentamicin (Table 2). Guanine plus cytosine content of DNA of 6 S. versatilis strains were in the range 38.3 to 40.1 mole% (Table 3).

Among the 11 strains of S. versatilis, 10 were gelatin negative but 7 were inulin positive. Only 2 strains grew anaerobically in the presence of fumarate. The results of 20 tests showing strain differences are summarized in Table 4.

Table 2. Similar characteristics of 11 strains of Sphingobacterium versatilis sp. nov.

| All 11 strains positive | All 11 strains negative |
|---|---|
| Slightly yellow pigment | Gram reaction, motility and fruity odor |
| Growth on BTB-lactose agar | Growth on Simmons citrate and NAC[b] agar |
| Growth in SMB[a] | Growth at 4C and 41C |
| Growth in 1% tryptone broth without NaCl | Growth in 1% tryptone broth cont. 5% NaCl |
| Growth in 1% tryptone Broth cont. 3% NaCl | Hydrogen sulfide in Kligler butt |
| Growth at 25C and 37C | Indole and acetylmethyl carbinol |
| Cytochrome oxidase and catalase | Nitrate reduction and nitrate respiration |
| Phosphatase production | L-lysine and L-ornithine decarboxylase |
| Hydrolysis of esculin | L-arginine dihydrolase |
| Ortho-nitro beta-D-galactopyranoside | Carlquist ninhydrin test and KCN resistance |
| Oxidative acid production from: | Glucose fermentation |
| D-arabinose and xylose | Utilization of acetate and malonate |
| Fructose, galactose and glucose | 2-ketogluconate and 3-ketolactonate |
| Mannose and melibiose | Acylamidase and hemolysis rabbit blood agar |
| Cellobiose, lactose, maltose, sucrose | Skim acetate agar, clear zone |
| Melezitose and raffinose | Oxidative acid production from: |
| Trehalose and salinine | Sorbitol, dulcitol and inositol |
| Susceptibility to clindamycin and minocyclin | Susceptibility to: |
| Sphingophospholipids in cellular lipids | Colistin, tobramycin and gentamicin |

a) Standard mineral base, Stanier et al. 1966, cont. 0.2% glucose and 0.1% ammon. sulfate.
b) Nalidixic acid-cetrimide agar, Goto and Enomoto 1970.

Table 3. Guanine plus cytosine mole percent of 6 strains of Sphingobacterium versatilis sp. nov.

| KM strain No. | Tm | GC mole%[a] |
|---|---|---|
| 926 | 85.5 | 39.6 |
| 1203 | 85.0 | 38.3 |
| 2087 | 85.8 | 40.1 |
| 2134 | 85.4 | 39.4 |
| 2138 | 85.7 | 39.9 |
| 2148 | 85.7 | 39.9 |

a) Average of 2 determinations.
Determined by K. Yamada, Ajinomoto Central Laboratory, Kawasaki.

The dendrogram (Fig. 1) based on the cluster analysis shows that the top cluster, composed of 11 strains of S. versatilis is clearly different from the 3 other major clusters. However, the top cluster is somewhat heterogeneous and should better be subdivided into 3 biovars. Type strain of C. johnsonae KM 2500 is related to the top cluster at 69 % similarity. The second and third clusters, Flavobacterium II b and F. meningosepticum, respectively, are clearly separated from each other and showing 90 % similarity within each cluster. The similarity of F. saccharolytica KM 1933 to the 3rd cluster is less than 80 %. F. odoratum cluster, the 2nd from the bottom, shows 83 % similarity to F. breve cluster at the bottom.

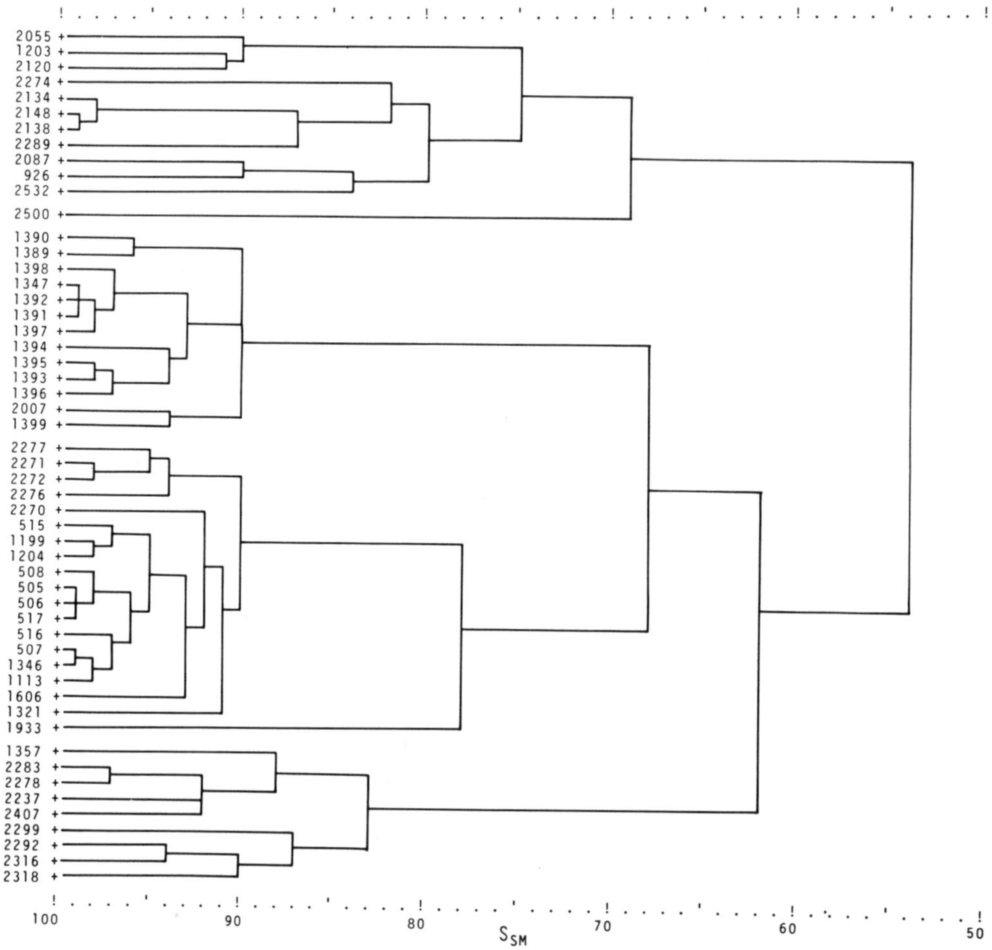


Fig. 1. A dendrogram based on the phenotypic characters of Sphingobacterium versatilis sp. nov., Cytophaga johnsonae and Flavobacterium species. Simple matching coefficients and the average linkage principle were used.

Based on the finding in dendrogram and the data summarized in Table 4, the 11 strains of S. versatilis were subdivided into 3 biovars (Table 5). All strains of biovar 1 and 2 were urease positive. Most of these strains were positive in extra-cellular DNase, acid production from inulin and hydrolysis of Tween 80. Most strains

Table 4. The 20 tests showing strain differences in the 11 strains of Sphingobacterium versatilis sp. nov.

| Substrate or test | Type strain | No. of strains positive | Strain No. which gave less common results |
|---|---|---|---|
| Growth on MacConkey agar | - | 1 | 2087 |
| Kligler slant, acid | + | 9 | 2120 |
| Citrate Christensen | - | 3 | 2274, 2289, 2532 |
| $H_2S$, Pb-acetate paper | + | 10 | 2274 |
| Nitrite to nitrogen | - | 2 | 1203, 2055 |
| Gelatin liquefaction | - | 1 | 2289 |
| Starch hydrolysis | - | 4 | 926, 2087, 2274, 2532 |
| Tween 80 hydrolysis | + | 7 | 1203, 2055, 2087, 2120 |
| Phenylalanine deaminase | - | 2 | 2055, 2532 |
| Acid from L-arabinose | - | 7 | 2134, 2138, 2148, 2274 |
| Acid from ethanol | - | 4 | 2134, 2138, 2148, 2289 |
| Acid from glycerol | + | 7 | 2055, 1203, 2087, 2120 |
| Acid from inulin | + | 7 | 1203, 2055, 2120, 2274 |
| Acid from mannitol | + | 4 | 2134, 2138, 2148, 2289 |
| Acid from rhamnose | + | 7 | 926, 1203, 2087, 2532 |
| Acid from D-ribose | - | 2 | 2055, 2120 |
| Green dyscoloration of rabbit blood agar | - | 5 | 926, 1203, 2055, 2120, 2289 |
| Sliding motility | + | 5 | 1203, 2120, 2134, 2138, 2148 |
| Anaerobic growth in fumarate broth | - | 2 | 2289, 2532 |
| Growth on 40% bile agar | + | 6 | 1203, 2055, 2087, 2120 |

Table 5. The 3 biovars among 11 strains of Sphingobacterium versatilis sp. nov.

| Substrate or test | Biovar and KM strain number | | | | | | | | | | |
|---|---|---|---|---|---|---|---|---|---|---|---|
| | 1 | | | | | 2 | | | 3 | | |
| | 2134 | 2138 | 2148 | 2274 | 2289 | 926 | 2087 | 2532 | 2120 | 1203 | 2055 |
| Urease | + | + | + | + | + | + | + | + | - | - | - |
| DNase | + | + | + | - | + | + | + | + | - | - | - |
| Acid from inulin | + | + | + | - | + | + | + | + | - | - | - |
| Hydrolysis of Tween 80 | + | + | + | + | + | + | - | + | - | - | - |
| Growth on 40% bile agar | + | + | + | + | + | + | - | + | - | - | - |
| $No_2$ to $N_2$ | - | - | - | - | - | - | - | - | - | + | + |
| Acid from ethanol | + | + | + | - | + | - | - | - | - | - | - |
| Acid from mannitol | + | + | + | - | + | - | - | - | - | - | - |
| Hydrolysis of starch | - | - | - | + | - | + | + | + | - | - | - |

of biovar 1 produced acid from ethanol and mannitol, and all strains of biovar 2 hydrolyzed starch. Three strains of biovar 3 were negative for all these tests but 2 reduced nitrite to nitrogen gas. A summary of the characteristics of S. versatilis compared with those of C. johnsonae, F. meningosepticum, Flavobacterium II b, and F. saccharolytica is shown in Table 6. S. versatilis can be clearly differentiated from the other 4 taxa.

As shown in Table 7, the major fatty acids of total extractable lipids of S. versatilis were 13-methyltetradecanoic (20 to 55%) and 2-hydroxy 13-methyltetradecanoic (11 to 34%) acids. Although the exact concentration of 15-methylhexadecanoic acid was not determined, the total percentage of branched chain acids exceeded 70% in 4 of 6 strains tabulated. The ceramide moiety of the sphingophospholipids in 6 strains of S. versatilis was composed of 2-hydroxy 13-methyltetradecanoic acid and branched chain C17 saturated long-chain base, specifically (Table 8). From the results of gas chromatography-mass spectrometry of TMS-ceramides, the most abundant species of ceramides was tentatively identified as N-2-hydroxy 13-methyltetradecanoyl-15(?)methylhexadecasphinganine (Fig. 2) /13, 26, 27, 28/.

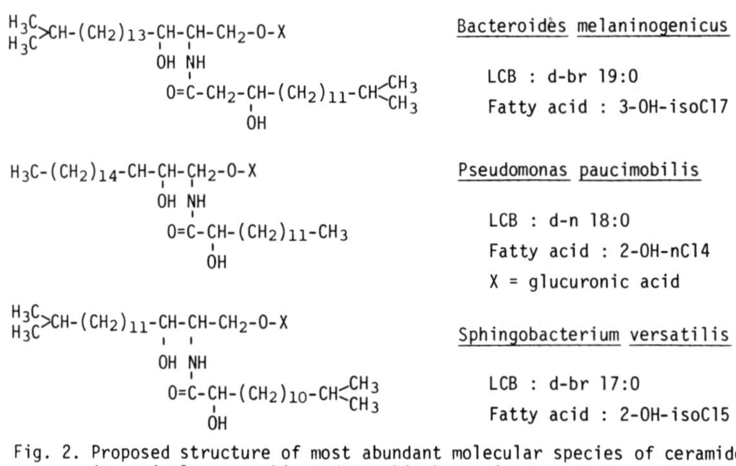


Fig. 2. Proposed structure of most abundant molecular species of ceramide in strictly anaerobic and aerobic bacteria

Classification and nomenclature of Flavobacterium / Cytophaga group of organisms are one of the blind spots in taxonomy of nonfermentative Gram-negative rods. Selection of taxonomic criteria is a difficult problem and each taxonomist might have a different opinion. In the classification and nomenclature of II k - 2 organism, we believe the presence of unique sphingophospholipids and the results of numerical analysis of phenotypic characters are important.

Table 6. Comparative characterization of Sphingobacterium versatilis sp. nov., Cytophaga johnsonae and 3 Flavobacterium species

| Substrate or test | Species (number of strains) | | | | |
|---|---|---|---|---|---|
| | Sphing. versatilis (11) | Cytophaga johnsonae (1) | Flavob. meningosept. (18) | Flavob. II b (13) | Flavob. saccharol. (1) |
| Deep yellow pigment | 0[a)] | + | 0 | 100 | - |
| Dark red in 20% KOH | 0 | + | 0 | 100 | - |
| Fruity odor | 0 | - | 0 | 100 | - |
| Indole | 0 | - | 100 | 100 | - |
| Acid from: | | | | | |
| D-arabinose | 100 | + | 100 | 0 | + |
| Xylose | 100 | + | 33 | 0 | - |
| Fructose and glucose | 100 | + | 100 | 100 | + |
| Galactose | 100 | + | 0 | 0 | + |
| Lactose | 100 | - | 100 | 0 | + |
| Maltose | 100 | + | 100 | 100 | + |
| Cellobiose | 100 | + | 100 | 0 | + |
| Sucrose | 100 | + | 0 | 0 | - |
| Melezitose | 100 | - | 0 | 0 | - |
| Salicin | 100 | + | 0 | 0 | - |
| Inulin | 64 | - | 0 | 0 | - |
| Fumarate, anaerobic | 19 | - | 89 | 100 | - |
| Malonate utilization | 0 | - | 0 | 100 | - |
| Egg yolk reaction | 0 | - | 0 | 100 | - |
| Skim acetate agar | 0 | + | 100 | 100 | + |
| Starch hydrolysis | 36 | + | 0 | 100 | + |
| Gelatinase | 9 | + | 100 | 100 | + |
| Urease | 73 | - | 0 | 0 | - |
| Growth in SMB contain. 0.2% glucose and 0.1% ammonium sulf. | 100 | + | 0 | 0 | - |

a) Number refers to percentage of positive reaction
\+ Positive reaction
\- Negative reaction

Description of Sphingobacterium gen. nov.: Gram-negative straight rod with rounded ends, occurring singly or in pairs. Nonmotile and nonfermentative. On heart infusion agar, pale yellow colonies with entire or irregular margin. The GC content of DNA is 38.9 ± 0.9 mole%. Cellular lipids contain sphingophospholipids. The most abundant molecular species of ceramide is N-2-hydroxy 13-methyltetradecanoyl-15(?)methylhexadecasphinganine. The predominant cellular fatty acid is 13-methyltetradecanoic acid, and 2-hydroxy 13-methyl-tetradecanoic acid is the next predominant. Other features of the genus are shown in Table 6. The strain KM 2138 was designated as the type strain for the species Sphingobacterium versatilis sp. nov.. A transfer of strain KM 2138 was deposited with the American Type Culture Collection on July 1, 1980. The accession

Table 7. Fatty acid composition of total extractable lipids of 6 strains of _Sphingobacterium versatilis_ sp. nov.

| Carbon number | KM strain number | | | | | |
|---|---|---|---|---|---|---|
| | 2134 | 2138 | 2148 | 926 | 2087 | 1203 |
| n 14 | 2.0 | 2.4 | 2.8 | 0.7 | 2.7 | 1.4 |
| i 15 | 38.7 | 41.0 | 30.2 | 26.4 | 21.6 | 55.6 |
| n 15 | 1.6 | 1.2 | 1.0 | 2.6 | 0.4 | 0.3 |
| i 16 | 1.1 | tr | tr | 1.4 | tr | tr |
| n 16 | 3.1 | 3.9 | 3.8 | 4.0 | 12.1 | 1.1 |
| 16:1 + i 17 | 13.4 | 12.6 | 15.2 | 19.8 | 31.1 | 20.7 |
| n 17 | 2.6 | 2.2 | 1.9 | 2.5 | tr | tr |
| $X_1$ | tr | tr | tr | 0.6 | tr | tr |
| $X_2$ | tr | tr | tr | tr | 2.4 | tr |
| 2-OH n 14 | tr | tr | tr | tr | 0.3 | tr |
| 2-OH i 15 | 27.2 | 27.4 | 34.5 | 32.7 | 28.5 | 11.1 |
| 3-OH n 14 | tr | tr | tr | 0.4 | tr | tr |
| 3-OH i 15 | 1.1 | 1.8 | 0.5 | 1.1 | 0.2 | tr |
| 3OH n 15 | 1.0 | tr | 1.5 | tr | 0.3 | tr |
| 3OH n 16 | tr | tr | tr | 0.8 | 0.3 | tr |
| 3-OH i, ai 17 | 7.7 | 7.4 | 8.7 | 7.1 | 0.4 | 9.0 |
| Total hydroxy | 37.0 | 36.6 | 45.2 | 42.1 | 29.7 | 20.1 |
| Total branched* | 75.8 | 77.6 | 73.9 | 68.7 | 50.7 | 76.3 |

n: Normal, i: Iso, ai: Anteiso, 2-OH: 2-hydroxy, 3-OH: 3-hydroxy,
X: Unknown, *: Iso 17 was not included.
Composition expressed as percentage.

Table 8. Long-chain base composition in the total lipids from 6 strains of _Sphingobacterium versatilis_ sp. nov.

| Long-chain base | KM strain number | | | | | |
|---|---|---|---|---|---|---|
| | 2134 | 2138 | 1248 | 926 | 2087 | 1203 |
| d-15:0 br | tr | tr | tr | tr | tr | tr |
| d-16:1 (?) | tr | tr | tr | 1.7 | tr | tr |
| d-16:0 | 4.2 | 2.8 | 4.8 | 4.5 | 3.6 | tr |
| d-17:0 br | 82.2 | 68.1 | 82.4 | 77.5 | 79.1 | 97.5 |
| d-18:1 (?) | 1.1 | tr | 1.6 | 2.3 | 2.3 | tr |
| d-18:0 | 12.5 | 25.3 | 11.2 | 14.0 | 15.0 | 2.5 |
| d-19:0 br | tr | 3.8 | tr | tr | tr | tr |

KM: Kansai Medical University, tr: Trace, br: Branched-chain.
Composition expressed as percentage.

number of the culture was announced to one of us (E.Y) as ATCC 33300, on July 15. Sphingobacterium gen. nov. is placed in the family Cytophagaceae. Type species of the genus is Sphingobacterium versatilis sp. nov..

Strains of Flavobacterium sp. group II b are regarded as conforming to a species.

## References


/1/ Callies, E., Mannheim, W.: Int. J. Syst. Bacteriol. 28, 14-19 (1978)

/2/ Christensen, P.J., Cook, F.D.: Can. J. Microbiol. 18, 1933-1940 (1972)

/3/ Fautz, E., Rosenfelder, G., Grotjahn, L.: J. Bacteriol. 140, 852-858 (1979)

/4/ Finnerty, W.R., Makula, R.A., Feeley, J.C.: Ann. Int. Med. 90, 631-634 (1979)

/5/ Flamm, H.: Zentralbl. Bakteriol. Abt. 1 Orig. 166, 398-502 (1956)

/6/ Folch, J., Lees, M., S.-Stanley, G.H.: J. Biol. Chem. 226, 497-509 (1959)

/7/ Fritsche, D., Boehmer, H.: Arch. Mikrobiol. 92, 359-363 (1973)

/8/ Gaver, R.C., Sweeley, C.C.: J. Amer. Oil Chem. Soc. 42, 294-298 (1965)

/9/ Henrichsen, J.: Bacteriol. Rev. 36, 478-503 (1972)

/10/ Holmes, B., Owen, R.J., Evans, A., Malnick, H., Willcox, W.R.: Int. J. Syst. Bacteriol. 27, 133-146.

/11/ Kunsman, J.E.: J. Bacteriol. 113, 1121-1126 (1973)

/12/ Labach, J.P., White, D.C.: J. Lipid Res. 10, 528-534 (1969)

/13/ Miyagawa, E., Azuma, R., Suto, T., Yano, I.: J. Biochem. 86, 311-320 (1979)

/14/ Mizuta, S., Imamura, K., Dateki, N., Suda, M., Kanetsugu, K., Tasaki, K.: J. Ped. Practice (in Japanese) 37, 1134-1140 (1974)

/15/ Moss, C.W., Dees, S.B.: J. Clin. Microbiol. 4:492-502 (1976)

/16/ Moss, C.W., Weaver, R.E., Dees, S.B., Cherry, W.B.: J. Clin. Microbiol. 6, 140-143 (1977)

/17/ Plackett, P., Smith, P.F., Mayberry, W.R.: J. Bacteriol. 104, 798-807 (1970)

/18/ Shaw, N., Stead, A.: Biochem. J. 143, 461-464 (1974)

/19/ Skerman, V.B.D., McGowan, V., Sneath, P.H.A. (eds.): Int. J. Syst. Bacteriol. 30, 225-420 (1980)

/20/ Steiner, S., Conti, S.F., Lester, R.L.: J. Bacteriol. 116, 1199-1211 (1973)

/21/ Sweeley, C.C., Moscatelli, E.A.: J. Lipid Res. 1, 40-47 (1959)

/22/ Tatum, H.W., Ewing, W.H., Weaver, R.E.: p. 270-294 in: Lennette, E.H., Spaulding, E.H., Truant, J.P. (eds.), Manual of clinical microbiol. 2nd ed. Washington D.C. (1974): Amer. Soc. Microbiol.

/23/ Tourtellotte, M.E., Jensen, R.G., Gander, G.W., Morowitz, H.J.: J. Bacteriol. 86, 370-379 (1963)

/24/ White, D.C., Tucker, A.N., Sweeley, C.C.: Biochim. Biophys. Acta 187, 527-532 (1969)

/25/ Yabuuchi, E., Tanimura, E., Ohyama, A., Yano, I., Yamamoto, A.: J. Gen. Appl. Microbiol. 25, 95-107 (1979)

/26/ Yamamoto, A., Yano, I., Masui, M., Yabuuchi, E.: J. Biochem. 83, 1213-1216 (1978)

/27/ Yano, I., Ohno, Y., Masui, M., Kato, K., Yabuuchi, E., Ohyama, A.: Lipids 11, 685-688 (1976)

/28/ Yano, I., Yamamoto, A., Ohno, Y., Masui, M., Yabuuchi, E.: Proc. Japan. Confer. Biochem. Lipids (in Japanese) 19, 19-22 (1977)

# A COMPARISON OF SOME GRAM-NEGATIVE, YELLOW PIGMENTED RODS ISOLATED FROM ENVIRONMENTAL AND CLINICAL SOURCES

K.S. Thomson[1] and T.A. McMeekin[2]

[1]Repatriation General Hospital, Hobart, Tasmania, 7000,
[2]Department of Agricultural Science, University of Tasmania, G.P.O. Box 252C, Hobart, Tasmania, 7001, Australia.

## Summary

A numerical taxonomic study was undertaken to compare 86 strains of Gram negative, yellow pigmented rods including 9 reference strains and 17 clinical isolates. Computation of data by a nearest neighbour sorting strategy separated a group of 28 strains at the 65% s level which corresponded to the revised genus description of *Flavobacterium* [12]. This group included strains of *F. breve* (NCTC 11099), *F. odoratum* (NCTC 11036), King's group IIb (NCTC 10795), clinical isolates and environmental isolates. *F. meningosepticum* (NCTC 10016), NCTC 10798 (King's group IIf) and *F. aquatile* (NCIB 8694) were excluded and the latter was contained in a group with *F. pectinovorum* (≡*C. johnsonae* NCIB 9059) and other gliding strains. In the *Flavobacterium* group four isolates were presumptively identified as *F. odoratum*, five isolates as *F. breve*, but the remainder could not be assigned to well defined species. The results of the study have been used to illustrate research needs for Gram negative yellow pigmented rods.

## Introduction

Since the preparation of material for the 8th edition of Bergey's Manual of Determinative Bacteriology [35], several taxonomic studies have been carried out on Gram negative, yellow pigmented rods from diverse sources (Table 1). These have provided much information on the relationships of flavobacteria and as a result a consensus about the taxonomic limits of the genus has emerged. Independently several workers [9, 12, 25] suggested that the genus be restricted to Gram negative, yellow pigmented rods which are non fermentative, non flagellate, do not spread or glide and have low % G + C ratios. This has led to a formal proposal that the genus description be emended and that the current type species *F. aquatile* (type strain ATCC 11947) be replaced by *F. breve* (neotype strain NCTC 11099) [12].

Table 1

Some Taxonomic Studies of Flavobacteria Since 1971.

| Authors | No. of Strains | Source of Isolates |
|---|---|---|
| Byrom [4] | 245 + 5 reference | Milk and canned products |
| McMeekin [22] | 69 + 10 reference | Poultry (42) |
| | | Meat (24) |
| | | Dairy (3) |
| Owen [28] | 93 + 20 reference | Human (53) |
| | | Hospital (12) |
| | | Miscellaneous (28) |
| Gavini and Leclerc [6] | 128 | Surface waters |
| Hayes [8] | 170 + 9 reference | Beef (29) |
| | | Lamb (28) |
| | | Pig (24) |
| | | Poultry (25) |
| | | Milk (25) |
| | | Water (28) |
| | | Soil (11) |
| Geringer and Kielwein [7] | 105 | Milk |

The taxonomic advances which have been made as a result of the accumulation and analysis of data from the above mentioned studies are however threatened by the report that *F. aquatile* (ATCC 11947) and *F. meningosepticum* (ATCC 13253) are flagellate organisms [33]. If substantiated this finding will result in a return to the taxonomic chaos traditionally associated with *Flavobacterium*.

During the seventies there has also been increased clinical interest in the genus and several workers have noted marked drug resistance patterns of Gram negative, yellow pigmented rods [1, 16, 30]. However, a feature of most taxonomic studies has been the inclusion of isolates from environmental or clinical sources but not from both.

This paper reports preliminary data from a numerical taxonomic study of isolates from both clinical and environmental sources and because of the serious taxonomic implications of the report of Webster and Hugh [33] the results of electron microscopic observations on *F. aquatile* and *F. meningosepticum* are also recorded.

## Materials and Methods

### BACTERIAL STRAINS

The origin of the 86 strains studied is shown in Table 2 and the identity of the 9 reference strains included is recorded in Table 3.

Table 2

Origin of Strains Included in the Study.

| Source | Number |
|---|---|
| Poultry processing water | 24 |
| Hospital water supply | 23 |
| Australian clinical isolates | 21 |
| Miscellaneous | 9 |
| Reference strains | 9 |
| Total | 86 |

Table 3

Reference Strains Included in the Study.

| Name | Culture Collection Number(s) |
|---|---|
| *F. aquatile* | NCIB 8694 (ATCC 11947) |
| *F. meningosepticum* | NCTC 10016 (ATCC 13253) |
| *F. pectinovorum* (≡*C. johnsonae*) | NCIB 9059 |
| *F. breve* | NCTC 11099 |
| *F. odoratum* | NCTC 11036 |
| *Flavobacterium* sp. King's Group IIb | NCTC 10795 |
| *Flavobacterium* sp. King's Group IIf | NCTC 10798 |
| *Pseudomonas paucimobilis* | NCTC 11030 |
| *P. paucimobilis* | CL162/70 (ex. B. Holmes NCTC) |

CHARACTERISATION TESTS

All strains were subjected to the characterisation tests listed in Table 4. Tests for susceptibility to antibacterial agents were performed on Wellcotest Sensitivity Test Agar (Wellcome) of 4mm depth. Inhibition zone diameters were measured using vernier calipers after 3 days incubation. Diameters equal to or greater than those listed in Table 5 were reported as positive. The zone diameters chosen do not have any clinical significance but were selected for their taxonomic potential on the basis of the distribution of resistance of the population to each antibacterial agent. Further tests to presumptively identify isolates as strains of *F. breve* or *F. odoratum* are listed in Table 7 and were performed as specified by Holmes *et al.* [14, 15].

COMPUTATION OF DATA

Computations were performed in a Cyber 76 computer using the Taxon package of taxonomic programmes. Similarity coefficients using the coefficient of Jaccard in which the negative matches were excluded were calculated and clustering was performed by the nearest and furthest neighbour sorting strategies.

Table 4

Cultural, Morphological, Physiological and Biochemical Characterisation Tests.

| Test | Method | Test | Method | Test | Method |
|---|---|---|---|---|---|
| Motility | (a) | Oxidase | (h) | Citrate utilisation | (o) |
| Gliding motility | (b) | DNAase | (i) | Esculin hydrolysis | (o) |
| Spreading growth | (b) | O-F test (glucose) | (j) | Acid from arabinose | (o) |
| Filament formation | (c) | Indole production | (k) | Acid from dextrose | (o) |
| Colony diameter - 72h | (d) | Phosphatase | (l) | Acid from dulcitol | (o) |
| Colonial appearance | (d) | Gelatin hydrolysis | (m) | Acid from galactose | (o) |
| Pigmentation | (d) | β-lactamase production | (n) | Acid from glycerol | (o) |
| Growth in nutrient broth | (c) | Urease | (o) | Acid from inositol | (o) |
| Growth at 37°C | (c) | $H_2S$ production | (o) | Acid from maltose | (o) |
| Growth on blood agar | (e) | Lysine decarboxylase | (o) | Acid from raffinose | (o) |
| Haemolysis | (e) | Ornithine decarboxylase | (o) | Acid from salicin | (o) |
| Growth on Oxoid MacConkey No. 3 | | Arginine dehydrolase | (o) | Acid from sorbitol | (o) |
| | | O.N.P.G. | (o) | Acid from sucrose | (o) |
| Growth in inorganic salts + glucose | (f) | Malonate utilisation | (o) | Acid from xylose | (o) |
| UV sensitivity | (g) | Voges Proskauer | (o) | | |

(a) wet mount and motility test medium (BBL); (b) Anacker and Ordal [2]; (c) nutrient broth (Oxoid); (d) nutrient agar (Oxoid); (e) Columbia Agar Base(Oxoid) + defibrinated horse blood (5% w/v); (f) Hayes [9]; (g) McMeekin [23]; (h) Kovacs [20]; (i) DNAase agar (Oxoid); (j) Hugh and Leifson [17]; (k) tryptone water (1% w/v); (l) Barber and Kuper [3]; (m) Charcoal gelatin discs (Oxoid), 7 days incubation; (n) O'Callaghan *et al.* [27]; (o) Minitek system (BBL).

Table 5

Antibacterial Agents and Inhibition Zone Diameters used in Characterisation Tests.

| Antibacterial agent | Concentration | Zone Diameter in mm | Antibacterial agent | Concentration | Zone Diameter in mm |
|---|---|---|---|---|---|
| Ampicillin | 25 mcg | 29 | Metronidazole | 5 mcg | any zone |
| Carbenicillin | 50 mcg | 27 | Nalidixic Acid | 30 mcg | 24 |
| Cephaloridine | 30 mcg | 20 | Neomycin | 30 mcg | 27 |
| Cephalothin | 30 mcg | 18 | Nitrofurantoin | 200 mcg | 20 |
| Chloramphenicol | 10 mcg | 20 | Novobiocin | 30 mcg | 18 |
| Clindamycin | 2 mcg | 28 | Penicillin | 1.5 units | 20 |
| Cloxacillin | 5 mcg | any zone | Polymyxin B | 300 units | 27 |
| Colistin | 10 mcg | 12 | Streptomycin | 10 mcg | 25 |
| Erythromycin | 10 mcg | 24 | Sulfafurazole | 100 mcg | 16 |
| Furacin | 100 mcg | 21 | Tetracycline | 10 mcg | 26 |
| Gentamicin | 10 mcg | 20 | Tobramycin | 10 mcg | 18 |
| Kanamycin | 30 mcg | 18 | Trimethoprim | 1.25 mcg | 19 |
| Lincomycin | 2 mcg | 20 | | | |

## Results

The nearest neighbour sorting strategy separated a group of 28 strains at the 65% s level which contained only organisms corresponding to the revised genus description of *Flavobacterium* [12]. The strains included in this group (Table 6) stand apart from gliding strains and other Gram negative, yellow pigmented rods such as *Pseudomonas paucimobilis*.

Table 6

Strains Contained in *Flavobacterium* Group.

| |
|---|
| *F. breve* NCTC 11099 |
| *F. odoratum* NCTC 11036 |
| King's Group IIb NCTC 10795 |
| 15 poultry plant chiller water isolates |
| 6 clinical isolates |
| 1 milk isolate |
| Q.684 ex. L. Sly, University of Queensland, Australia |
| Strain 015 ex. J.M. Shewan from Metal Box Co. Ltd. |
| Strain 25 bulk milk tank swab [24, 26] |

The nearest neighbour sorting strategy used is the least discriminatory sorting strategy available and although separation of isolates conforming to the revised genus description was indicated,little useful information was provided about the relationship of organisms within this group. Preliminary data from the furthest neighbour sorting strategy indicates that the reference strains *F. breve* (NCTC 11099), *F. odoratum* (NCTC 11036) and NCTC 10795 (King's Group IIb) occur in separate groups but many other organisms, particularly isolates from environmental sources, are distinct from the marker strains.

In an attempt to presumptively identify strains to species level, organisms corresponding to the revised genus description were compared with the salient features listed for *F. breve* and *F. odoratum* [14, 15]. Three strains from clinical sources and one water isolate were presumptively identified as *F. odoratum* (Table 7). Strain U59 differed from NCTC 11036 in being gelatin negative and strain R37 showed a positive lysine decarboxylase reaction. To our knowledge strain U59, if confirmed as *F. odoratum*, is the first isolate of this species from a non clinical environment.

Five strains were presumptively identified as *F. breve* (Table 7). Strain U31 gave the same reactions as NCTC 11099 in this restricted series of tests, as did strain U37 except that the latter did not grow at 37°C. Similarly strains U1 and U10 were 37°C negative and like ATCC 14234 did not liquefy gelatin. Strain R26 was similar to NCTC 11099 in most respects but reduced nitrite, a feature associated with *F. odoratum* but not *F. breve* [14, 15].

Table 7

Some Characteristics of Strains Presumptively Identified as *F. odoratum* or *F. breve*.

| Test | *F. odoratum* NCTC 11036* | U59 | R25 | R28 | R37 | *F. breve* NCTC 11099* | NCTC 14234* | U1 | U10 | U31 | U37 | R26 | U38 |
|---|---|---|---|---|---|---|---|---|---|---|---|---|---|
| ASS glucose acid | - | - | - | - | - | + | + | + | + | + | + | + | - |
| ASS maltose acid | - | - | - | - | - | + | + | - | + | + | + | + | - |
| DNAase | + | + | + | + | + | + | + | + | + | + | + | + | + |
| Gelatin liquefaction | + | - | + | + | + | + | - | - | - | + | + | + | - |
| Indole production† | - | - | - | - | - | + | + | + | + | + | + | + | + |
| Lysine decarboxylase | - | - | - | - | + | - | - | - | - | - | - | - | - |
| Nitrite reduction | + | + | + | + | + | - | - | - | - | - | - | + | + |
| O-F alkaline | + | + | + | + | + | - | - | - | - | - | - | - | + |
| Urease production | + | + | + | + | + | - | - | - | - | - | - | - | - |
| Growth at 37°C | + | + | + | + | + | + | - | - | - | + | - | + | - |

*Date of Holmes *et al.* [14, 15]. †Ehrlich reagent.

Strain U38 presents particular taxonomic problems and shares a number of features usually associated with either *F. odoratum* or *F. breve* (Table 7). Lack of acid production from ASS glucose and maltose, nitrite reduction and an alkaline O-F reaction suggest affinity with *F. odoratum*, whereas indole production and a negative urease reaction are consistent with *F. breve*.

## Discussion

### THE REVISED GENUS *FLAVOBACTERIUM*

Although a consensus opinion may have arisen about the taxonomic limits of the genus *Flavobacterium*, differentiation within the revised genus remains far from satisfactory. At this time only three species (*F. meningosepticum, F. breve, F. odoratum*) have been adequately described, and these are based almost solely on descriptions of clinical isolates.

This study has presumptively identified further strains of *F. breve* and *F. odoratum* from both clinical and environmental sources but many other isolates could not be assigned to species. The taxonomic problem is exemplified by strain U38 and may foreshadow the difficulties of allocating many 37°C negative isolates to particular species. Such a situation is perhaps not unexpected and might be inferred from the results of several numerical taxonomy exercises. In these it has been possible to delineate a group of Gram negative, yellow pigmented rods corresponding to the present concept of *Flavobacterium* but within which distinct phenons possibly corresponding to species have not been evident .

This does not negate the contention of Holmes *et al.* [14, 15] that *F. breve* and *F. odoratum* are distinct species nor does it detract from the value of having well defined reference strains as a framework with which to compare new isolates. However, it should be noted that the great majority of strains studied by these authors were isolated from clinical sources where environmental pressures, particularly the ability to grow at 37°C and possibly other selection pressures might be expected to lead to the recovery of more discrete species.

This point can only be resolved by further studies of environmental isolates. The data is urgently required as due to the relative taxonomic activity of workers interested in environmental isolates, the genus has moved inexorably in a "growth at 37°C positive, clinical direction", whilst the great majority of isolates do not grow at 37°C and are recovered from non clinical situations. In particular the status of *F. breve* as an "environmental species" requires elucidation. A strain of this organism (NCTC 11099) has been proposed to replace *F. aquatile* as type species of the genus [12]. The neotype of *F. breve* was chosen in preference to ATCC 14234, the only non clinical, 37°C negative strain examined. ATCC 14234 was already held in a culture collection but it did not originate from the original author of the species although it conformed to the original meagre description [15]. It was the least typical of the 7 strains examined and therefore was not considered a suitable neotype strain. However, in this study a further three, 37°C negative strains have been isolated which have been presumptively identified as *F. breve* and which may increase the "typical" status of ATCC 14234.

## *FLAVOBACTERIUM AQUATILE* AND *F. MENINGOSEPTICUM*

Two reference species listed by Weeks [35] in Section 1 *Flavobacterium* were excluded from the group corresponding to the revised genus description - these were *F. aquatile* and *F. meningosepticum*.

The taxonomic problems surrounding *F. aquatile* have been well documented [12, 25, 31] and previous numerical taxonomic studies have also shown ATCC 11947 to cluster separately from other Gram negative, yellow pigmented rods [9]. This study has confirmed the lack of phenetic similarity between *F. aquatile* and the majority of *Flavobacterium* isolates from both clinical and environmental sources and adds support to the proposal of Holmes and Owen [12] that *F. aquatile* be replaced as type species of the genus. It is also of interest that in both sorting strategies used *F. aquatile* was contained in groups which also included gliding isolates.

Although *F. meningosepticum* adheres to the revised genus description and has been maintained as a valid species by Holmes and Owen [12], it failed to cluster with any other strain. Similarly, McMeekin *et al.* [26] found, using a single linkage sorting strategy, that *F. meningosepticum* appeared as a singleton. The relationship of this organism to other flavobacteria, particularly those of clinical interest, warrants further consideration.

The problems of motility and flagella arrangement in high % G + C,yellow pigmented Gram negative rods have been well documented [13, 36]. However, until the report of Webster and Hugh [33] low % G + C Gram negative, yellow pigmented rods had been assumed to be aflagellate. In studies to be reported in full elsewhere, electron microscopy was used to examine the same strains of *F. meningosepticum* and *F. aquatile* for the presence of flagella. No flagella were found on either, but the latter displayed strands similar to those described as pseudoflagella by Weeks [34] and on myxobacteria [5, 18]. Follett and Webley [5] also noted "a remarkable and unexpected similarity between this organism and *C. johnsonae* (syn. *F. pectinovorum*) which has also been shown to display aberrant surface translocation characteristics [21, 24].

## DIFFERENTIATION FROM OTHER GRAM NEGATIVE, YELLOW PIGMENTED RODS

Organisms corresponding to the revised genus description of *Flavobacterium* have been shown by several numerical taxonomic studies to be distinct from gliding strains, but there is no easily performed test, apart from the demonstration of gliding motility which can be used to separate *Flavobacterium* from *Flexibacter/ Cytophaga*. Long flexuous cells and ability to grow in an $NH_4$ salts/glucose medium may also be useful indicators of *Flexibacter/Cytophaga* [10] but require to be confirmed by demonstration of gliding motility. This study has confirmed the separation of gliders and non gliders but no additional distinctive features have emerged. Further work is therefore required on the differentiation of *Flavobacterium* from low % G + C gliding organisms.

Current taxonomic opinion favours retention of the low % G + C non gliders as *Flavobacterium* but an alternative, rejected by McMeekin and Shewan [25], would be to regard these organisms as atypical flexibacters or cytophagas which have a deficiency in the mechanism of gliding motility. The report of Pate *et al.* [29],in which bacteriophage resistant, non gliding mutants of *C. johnsonae* are described, adds a further fascinating facet to the debate. How can these organisms be separated from *Flavobacterium*? Are organisms adhering to the revised genus description also resistant to similar bacteriophages? If so, can reversion to bacteriophage sensitive, gliding forms be induced? Positive answers to the latter two questions would strengthen the suggestion that these organisms should be regarded as atypical cytophagas or flexibacters, but this does not necessarily remove the requirement for a taxonomic unit in which to accommodate the revised genus *Flavobacterium*. McMeekin and Shewan [25] pointed to the difficulties of accommodating non gliding organisms in the *Myxobacterales* and in any case the distinction between gliders and non gliders in phenetic analyses [9, 26] suggests many other differences in addition to gliding motility and spreading growth.

Retention of the generic title *Flavobacterium* for non gliding low % G + C strains leaves the high % G + C in taxonomic limbo although it seems possible that these might eventually be accommodated in the genus *Empedobacter* [9,25]. A serious problem which restricts work on the high % G + C group is the small number of strains which are available for comparative studies. In this study, as in many others, the only high

% G + C strains isolated (as judged by UV sensitivity) were polar flagellate and allocated to *Pseudomonas*. In order to obtain additional isolates it may be possible to develop a selective isolation procedure based on UV light sensitivity as the high % G + C strains have been reported to be at least 1000 times more resistant to UV damage than low % G + C strains [23]. The search for additional features which may be correlated with % G + C should also continue and in this context bacterial sensitivity to X-rays and nitrogen mustard is correlated with % G + C content [11,19].

## Conclusions

Whilst considerable data has been accumulated on the biology of Gram negative, yellow pigmented rods, the need for additional work is evident and some suggestions have been made above. If the redefined genus *Flavobacterium* is to become a taxonomic reality a research priority is the description of further species within the genus. At present only 3 species are adequately described and all are based largely on clinical isolates. There is a need for a balance to be struck with detailed descriptions of environmental strains and in this context it will be of interest to see if further discrete species can be described. In addition, the report of Pate *et al.* [29] should provide the impetus for further work on the relationship between the revised genus *Flavobacterium* and *Flexibacter/Cytophaga*.

## References

[1] Altman, G., Bogoskovsky, B.: J. Med. Microbiol. 4, 296-299 (1971).

[2] Anacker, R.L., Ordal, E.J.: J. Bacteriol. 78, 25-32 (1959).

[3] Barber, M., Kuper, S.W.A.: J. Path. Bacteriol. 63, 65-68 (1951).

[4] Byrom, N.A.: J. Appl. Bacteriol. 34, 339-346 (1971).

[5] Follett, E.A.C., Webley, D.M.: Antonie van Leeuwenhoek, 31, 361-382 (1965).

[6] Gavini, F., Leclerc, H.: Revue Intl. Ocean. Med. 37/38, 17-68 (1975).

[7] Geringer, M., Kielwein, G.: Arch. Lebensmittelhyg. 27, 11-17 (1976).

[8] Hayes, P.R.: Ph.D. Thesis, University of Leeds (1975).

[9] Hayes, P.R.: J. Appl. Bacteriol. 43, 345-367 (1977).

[10] Hayes, P.R., McMeekin, T.A., Shewan, J.M.: in Skinner, F.A., Lovelock, D.W. (eds.). Identification Methods for Microbiologists, 2nd ed. London (1979): Academic Press.

[11] Haynes, R.H., Alexrod, D.E.: J. Cell Biol. 19, 33A (1963).

[12] Holmes, B., Owen, R.J.: Intl. J. Syst. Bacteriol. 29, 416-426 (1979).

[13] Holmes, B., Owen, R.J., Evans, A., Malnick, H., Willcox, W.R.: Intl. J. Syst. Bacteriol. 27, 133-146 (1977).

[14] Holmes, B., Snell, J.J.S., Lapage, S.P.: Intl. J. Syst. Bacteriol. 27, 330-336 (1977).

[15] Holmes, B., Snell, J.J.S., Lapage, S.P.: Intl. J. Syst. Bacteriol. 28, 201-228 (1978).

[16] Holmes, B., Snell, J.J.S., Lapage, S.P.: J. Clin. Pathol. 32, 73-73 (1979).

[17] Hugh, R., Leifson, E.: J. Bacteriol. 66, 24-26 (1953).

[18] Humphrey, B.A., Dickson, M.R., Marshall, K.C.: Arch. Microbiol. 120, 231-238 (1979).

[19] Kaplan, H.S., Zavarine, R.: Biochem. Biophys. Res. Commun. 8, 496-503 (1961).

[20] Kovacs, N.: Nature, Lond. 178, 703 (1956).

[21] Lund, B.M.: J. Appl. Bacteriol. 32, 60-67 (1969).

[22] McMeekin, T.A.: Ph.D. Thesis, Queen's University, Belfast (1971).

[23] McMeekin, T.A.: J. Gen. Microbiol. 103, 149-151 (1977).

[24] McMeekin, T.A., Patterson, J.T., Murray, J.G.: J. Appl. Bacteriol. 34, 699-716 (1971)

[25] McMeekin, T.A., Shewan, J.M.: J. Appl. Bacteriol. 45, 321-333 (1978).

[26] McMeekin, T.A., Stewart, D.B., Murray, J.G.: J. Appl. Bacteriol. 35, 129-137 (1972).

[27] O'Callaghan, C.H., Morris, A., Kirby, S., Shingler, A.H.: Antimicrobial Agents and Chemotherapy 1, 283-288 (1972).

[28] Owen, R.J.: M.Phil. Thesis, University of London (1972).

[29] Pate, J.L., Petzold, S.J., Chang, E.L.Y.: Curr. Microbiol. 2, 257-262 (1979).

[30] Pedersen, B.A., Marso, E., Pickett, M.J.: Amer. J. Clin. Pathol. 51, 178-192 (1970).

[31] Perry, L.B.: J. Appl. Bacteriol. 36, 227-232 (1973).

[32] Stutzer, M., Kwaschnina, A.: Zentrabl. Bakteriol. Parasitenkd. Infektionskr. Hyg. Abt. I. Orig. 113, 219-225 (1929).

[33] Webster, J.A., Hugh, R.: Intl. J. Syst. Bacteriol. 29, 333-338 (1979).

[34] Weeks, O.B.: J. Bacteriol. 69, 649-658 (1955).

[35] Weeks, O.B.: p. 357-363 in: Buchanan, R.E., Gibbons, N.E. (eds.): Bergey's Manual of Determinative Bacteriology, 8th edn. Baltimore (1974): Williams and Wilkins Co.

[36] Yabuuchi, E., Tanimura, E., Ohyama, A., Yano, I., Yamomoto, A.: J. Gen. Appl. Microbiol. 25, 95-108 (1979).

# THE FLEXIRUBIN-TYPE PIGMENTS, CHEMOSYSTEMATICALLY USEFUL COMPOUNDS

H. Reichenbach[1], W. Kohl[2], H. Achenbach[2]

[1]Gesellschaft für Biotechnologische Forschung,
Mascheroder Weg 1, D-3300 Braunschweig-Stöckheim, and
[2]Chemisches Laboratorium der Universität,
Albertstr. 21, D-7800 Freiburg, FRG

## Summary

The flexirubin-type pigments are compounds of a novel chemical structure: A polyenoic acid chromophore with a *para*-hydroxy phenyl ring (ring A) in the end position is esterified with a di-alkylated resorcinol (ring B). The basic structure may be modified by a) a variation of the chain length of the chromophore between 6 and 8 conjugated double bonds; b) a methylation in the *meta*-position in ring A; c) a chlorination in the *meta*-position in ring A; and d) broad variations of the hydrocarbon chains in ring B with respect to the chain length and linearity. All strains, which contain flexirubin-type pigments, also produce chlorinated counterparts of most of the structural variants. The distribution of flexirubin-type pigments among yellow bacteria makes them very useful chemosystematic markers. They are found in most strains of *Flexibacter* and *Cytophaga*-like bacteria isolated from soil and freshwater, in *Sporocytophaga*, and in many flavobacteria of the low GC-group. They seem to be absent in gram-positive bacteria, in flagellated bacteria, and in most *Cytophaga*-like bacteria isolated from marine environments. Many bacteria, which produce flexirubin-type pigments, contain complex mixtures of up to 25 different structural variants, but nevertheless it seems that each taxonomic group synthesizes its own specific modification of the pigment. In contrast to carotenoids, also present in many of the above-mentioned organisms, flexirubin-type pigments are located in the outer membrane of the gram-negative cell wall. The biological function of the new pigments is still obscure: Unpigmented mutants can be obtained, and grow well under laboratory conditions.

Until a few years ago it was generally assumed that the pigments of yellow gliding bacteria are always carotenoids /15, 20, 21, 26/. When we started to analyze the pigment patterns of Flexibacter elegans and of several Cytophaga-like bacteria, it soon became apparent, however, that at least some of the pigments produced by these organisms could not be carotenoids /23, 24/. This conclusion was based on three different observations: a) The electron spectra of the pigments did not show any fine structure (Fig. 1). With carotenoids, "degenerate" spectra like the recorded ones are usually due to conjugated keto-carbonyl functions, but such carbonyl functions could not be demonstrated in the new pigments. b) There was a large and reversible shift in light absorption to longer wavelengths, when the pH of alcoholic solutions was raised to alkaline conditions (Fig. 1). Absorption shifts of this type are very rare with carotenoids, and actually restricted to certain phenolic compounds. c) Radioactive mevalonic acid was not incorporated into the new pigments, although the menaquinones of the organisms became heavily labelled.

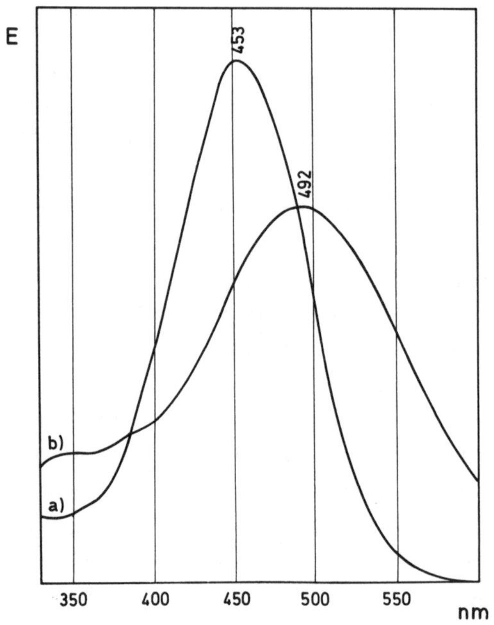


Fig. 1. Electron spectra of flexirubin from Flexibacter elegans, a) in ethanol, b) in alkaline ethanol

We, therefore, decided to work out the complete chemical structure of the said pigments, and chose Flexibacter elegans, strain Fx el, as the first example. As it turned out later, this choice was a very lucky one, because this bacterium produces essentially only two pigments, while most other organisms contain extremely complicated mixtures of up to 25 closely related structures.

The pigments of Fx. elegans Fx el were indeed of a novel type, and the main compound was named flexirubin /3, 10, 24/ (Fig. 2). Later we could demonstrate, that flexirubin-type pigments are very common in certain groups of bacteria, and up till now about 50 different structures have been elucidated.

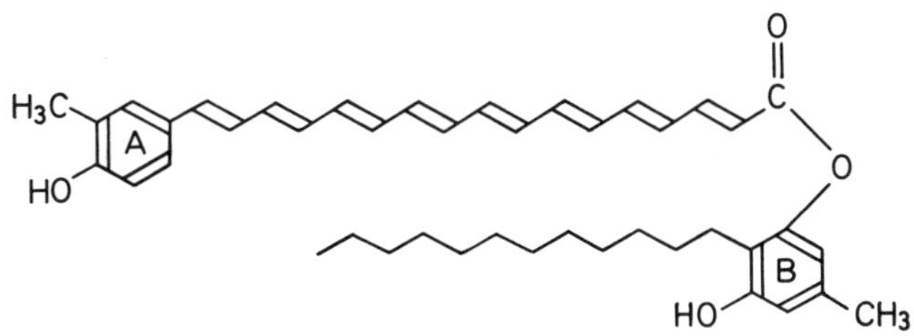


Fig. 2. Flexirubin from Flexibacter elegans.

The following structural elements are characteristic of flexirubin-type pigments (Figs. 2, 3) /5/: a) A polyenoic acid chromophore with a para-hydroxy phenyl ring (ring A) in the omega-position. b) A resorcinol (ring B) substituted with two hydrocarbon chains in the 2' and 5' position, and connected to the carboxyl group of the polyenoic acid via an ester bond.

Different flexirubin-type pigments arise by variation of the following structural elements /3, 4, 5, 6, 8, 9/: a) The length of the chromophore may vary between 6 and 8 double bonds; the main compounds often contain octaenoic acid. b) There may or may not be a methyl group in the m-position of ring A. c) The alkyl substituents in ring B may vary within wide limits with respect to chain length and chain branching. There never seem to be identical substituents in the 2' and 5' position of a resorcinol. The longest chains observed so far consist of 12 C-atoms. d) All strains investigated to date synthesize a chlorinated counterpart for most structural variants of the flexirubin type-pigment they contain. The chlorine is introduced into the m-position of ring A. Occasionally there are also dichloro-pigments present.

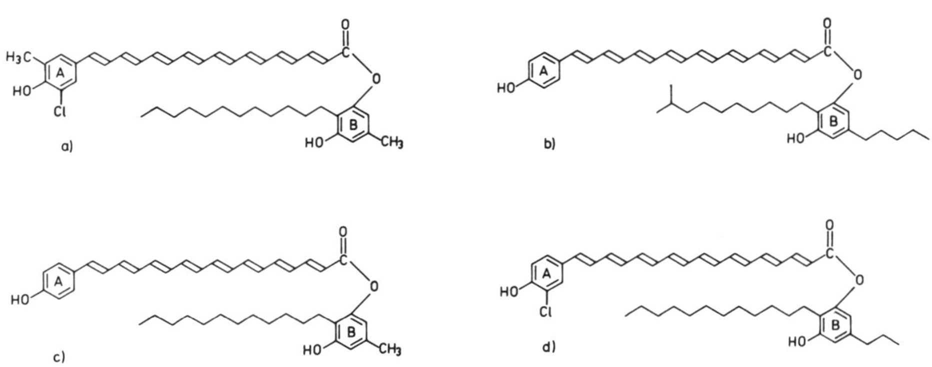


Fig. 3. Flexirubin-type pigments. a) Chloroflexirubin from Flexibacter elegans Fx el; b) One of the pigments of Cytophaga johnsonae Cy jl; c) and d), two examples from Flavobacterium strain C 1/2 (from Dr. Hayes, Leeds).

Biosynthetically, the precursor of ring A and of the adjacent three C-atoms of the chromophore is tyrosine, at least with Fx. elegans Fx el /7, 13/. The rest of the polyenoic chain originates from acetate. It seems that the complete substituted resorcinol moiety is ultimately derived from acetate, too, but the immediate biosynthetic precursors are not yet known with certainty. There may be two fatty acids involved. If a methyl group is present in Ring A, it is introduced from methionine.

The presence of flexirubin-type pigments in an organism can be deduced from the following observations: a) When an agar culture (or a pellet of cells harvested from a liquid culture) is covered with a 20 % KOH solution, the yellow or orange color of the colonies (pellet) shifts to red, purple, or brown. If the alkali is replaced by hydrochloric acid, the yellow color of the cell mass returns. b) The pigments can be extracted with acetone from the wet cell mass and purified by chromatography on silica gel thin layers. With the solvent system: petroleum ether (b.p, 60-80$^{0}$)/toluene/acetone = 25:25:16, $R_f$-values between 0.35 and 0.75 are to be expected. The phenolate reaction mentioned above can be performed directly on the thin layer plate by dotting the pigment spots with alkali and acid. c) The separated pigments can be eluted from the silica gel with acetone and their absorption spectra recorded. Typical spectra are often obtained even when the bands still consist of several pigments, because either the chromophores of the individual pigments are identical, or one of the pigments is present in a large excess. If the solvent is ethanol or methanol, the characteristic spectral shift can be produced directly in the measuring cuvette by adding a trace of alkali or acid. d) The Carr-Price reaction may also be performed on the thin layer plate: With a drop of a saturated solution of $SbCl_3$ in chloroform the pigments turn sky blue, with a drop of concentrated sulfuric acid deep violet blue. e) The most specific test for flexirubin-type pigments available at present is the addition of suitable radioactive precursors of the pigments to liquid cultures, and the demonstration of their incorporation into the pigments by autoradiography of the chromatograms /14/. This experiment can be done with culture volumes as small as 2 ml and is thus well suited for screening purposes. By applying labelled tyrosine, methionine, or mevalonic acid, respectively, to different batches, it is usually possible to differentiate between flexirubin-type pigments and carotenoids, and between flexirubin-type pigments with and without a methyl group in ring A. f) However, it must be emphasized that none of the above-mentioned tests really proves the presence of flexirubin-type pigments, and only a combination of several observations makes the conclusion reasonably reliable. Certainty can be obtained only by chemical characterization of the pigments, particularly by mass and NMR spectroscopy /1/. New structural variants are still to be expected and could be taxonomically interesting, but unfortunately the elucidation of the pigment pattern of a new strain is a very time-consuming process because of the difficult separation problems involved.

The biological function of the novel pigments is not understood at the moment. For several reasons it seems likely that flexirubin-type pigments are an essential component of the bacterial cell: a) The pigments have a relatively complicated chemical structure. Their biosynthesis therefore requires several specific enzymes, which would probably have been eliminated, if there had not been some kind of selection pressure for their conservation. b) While we isolate our strains ordinarily under conditions that do not allow us to recognize pigmentation, we virtually never get organisms that later upon cultivation on suitable media turn out to be unpigmented; i.e., in the natural environment there seems to be a strong preference for pigmented strains. c) The novel pigments are not typical "waste" products due to side reactions or imbalances in metabolism, for their biosynthesis is strictly correlated with growth: resting cells do not produce flexirubin-type pigments, nor is there growth without pigment synthesis /24/.

In contrast to carotenoids, the function of flexirubin-type pigments seems not to be protection against photo-oxidation: unpigmented mutants of Fx. elegans are as resistant to irradiation with light intensities up to 100 000 lux as is the pigmented wild strain (unpublished results). In fact, many strains contain carotenoids in addition to flexirubin-type pigments, which would be meaningless if both could serve the same function. Mutants, which do not produce flexirubin-type pigments any longer, can be readily isolated and grow very well under laboratory conditions. So far no clear differences in the behavior of such mutants and that of the wild strain could be detected.

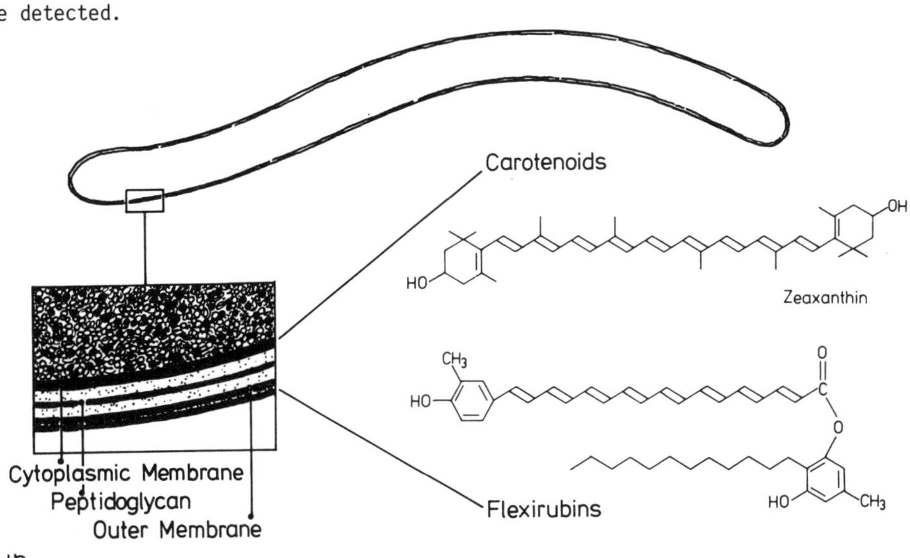


Fig. 4. Intracellular location of flexirubin and of carotenoids in Flexibacter elegans Fx 3/4 /17/.

As a first approach to the problem of the biological function, the intracellular location of flexirubin-type pigments was determined: Unexpectedly, the novel pigments are found in the outer membrane of the gram-negative cell wall /17/. If, in addition, there are carotenoids present in the cell, their location is the cytoplasmic membrane, as is known from other bacteria, and the two types of pigments can be separated from one another by separating the two membranes (Fig. 4). The fact, that flexirubin-type pigments are deposited in the cell wall, could explain, why the biosynthesis of the pigments is restricted to growing cultures, and why these compounds cannot have a photoprotective function. On the other hand, it gives rise to interesting questions as to the site of pigment synthesis and the mechanism of translocation from within the cell.

The distribution of flexirubin-type pigments among bacteria is not yet fully known. The novel compounds seem to be restricted to certain taxonomic groups, so that they can be used as chemosystematic markers /29, and unpublished results/. Flexirubin-type pigments seem to be characteristic for many Cytophaga-like bacteria from soil and freshwater, including the cellulose-decomposing cytophagas, Flexibacter, and Sporocytophaga. They have also been discovered in immotile flavobacteria with a low GC-content /9, 25/, and this occurrence may point to a relatively close taxonomic relationship between the two groups of bacteria. Flexirubin-type pigments are found only rarely with Cytophaga-like bacteria isolated from marine environments, which probably means, that cytophagas from marine and from terrestrial habitats are not very closely related, and that there is little exchange between the two populations even where they are in direct contact, as in coastal areas. The novel pigments are definitely absent from myxobacteria, Lysobacter, Herpetosiphon, Saprospira, and several other gliding bacteria. They have never been found with any gram-positive bacterium, nor with any flagellated gram-negative bacterium. There is also no known example of a eukaryotic organism containing flexirubin-type pigments.

It is interesting to note that each of the taxonomic groups mentioned above seems to produce its own modification of flexirubin-type pigments (Fig. 3): Thus only the Flexibacter pigments have a methyl group in ring A (and therefore become labelled with $^{14}CH_3$-)-methionine). The Cytophaga pigments have a large variety of different hydrocarbon chains as substituents in ring B, but never a methyl group as in the Flexibacter and in some of the Flavobacterium pigments. The Flavobacterium pigments, like the Cytophaga pigments, have no methyl group in ring A. They may have quite different substituents in ring B, but unlike all Cytophaga pigments known so far they may also contain the true flexirubin resorcinol with a methyl and a $C_{12}H_{25}$ substituent. The latter substitution pattern is the only one found in Flexibacter pigments. Of course, it is too early to say whether or not these rules are really general, because only a few strains have been analyzed in full detail. It would be of particular interest to see which variants of flexirubin-type pigments are produced by the cellulose-decomposing cytophagas, by Sporocytophaga, and by the diverse flavobacteria.

While the type of pigments characterized above has only been found in a rather restricted group of bacteria, molecules more or less related to flexirubin-type pigments and their structural elements have been described for quite different organisms (Fig. 5). Thus similar polyenoic acid chromophores with a phenyl ring in omega-position are known from cortisalin produced by the fungus Corticium salicinum /16/, and from the Xanthomonas pigments /11, 12/. Remotely related are asperenone isolated from the molds Aspergillus niger and A. awamori /22/, and the navenones from the marine slug, Navanax inermis /27/ Dialkylresorcinols related to those found in flexirubin-type pigments, have been described from the hyphomycete Stemphylium majusculum /28/, and from Pseudomonas sp. /19/. The latter compound was discovered through its antibiotic activity. A comparison of several synthetic resorcinols with different substitution patterns demonstrated, that the length of the attached hydrocarbon chains may be modified only within narrow limits, if the inhibitory effect of the compound is not to be lost /2/.

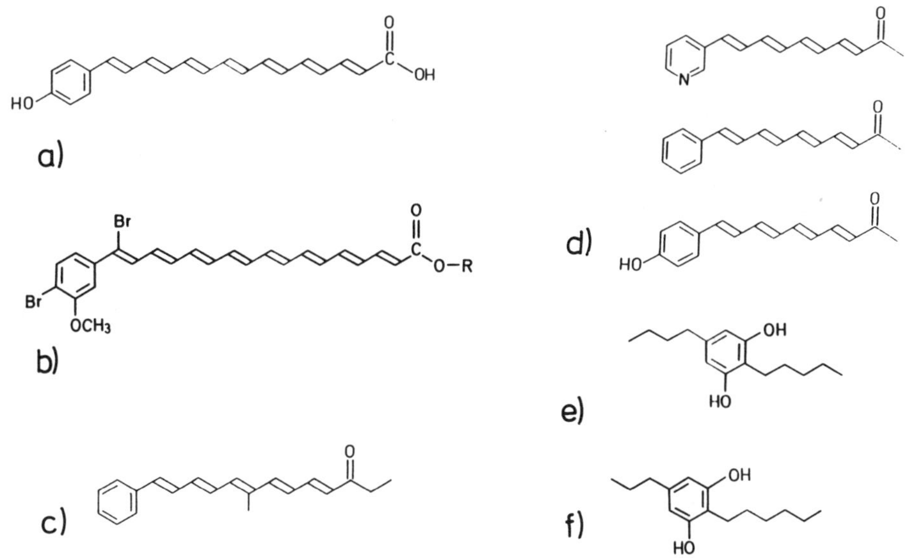


Fig. 5. Compounds structurally related to flexirubin-type pigments. a) Cortisalin; b) xanthomonadin I; c) asperenone; d) navenones A, B and C; e) stemphol (or its structural isomer); f) antibiotic DB-2073.

## References


/1/ Achenbach, H., Kohl, W.: Chem. Ber. 112, 209-217 (1979)

/2/ Achenbach, H., Kohl, W., Kunze, B.: Chem. Ber. 112, 1841-1848 (1979)

/3/ Achenbach, H., Kohl, W., Reichenbach, H.: Chem. Ber. 109, 2490-2502 (1976)

/4/ Achenbach, H., Kohl, W., Reichenbach, H.: Liebigs Ann. Chem. 1977, 1-7

/5/ Achenbach, H., Kohl, W., Reichenbach, H.: Rev. Latinoam. Quim. 9, 111.124 (1978)

/6/ Achenbach, H., Kohl, W., Reichenbach, H.: Chem. Ber. 112, 1999-2011 (1979)

/7/ Achenbach, H., Böttger, A., Kohl, W., Fautz, E., Reichenbach, H.: Phytochem. 18, 961-963 (1979)

/8/ Achenbach, H., Kohl, W., Alexanian, S., Reichenbach, H.: Chem. Ber. 112, 196-208 (1979)

/9/ Achenbach, H., Kohl, W., Böttger-Vetter, A., Reichenbach, H.: Tetrahedron, in print

/10/ Achenbach, H., Kohl, W., Reichenbach, H., Kleinig, H.: Tetradr. Lett. 1974, 2555-2556

/11/ Andrewes, A.G., Hertzberg, S., Liaaen-Jensen, S., Starr, M.P.: Acta Chem. Scand. 27, 2383-2395 (1973)

/12/ Andrewes, A.G., Jenkins, C.L., Starr, M.P., Shepherd, J., Hope, H.: Tetrahedr. Lett. 1976, 4023-4024

/13/ Fautz, E., Reichenbach, H.: Phytochem. 18, 957-959 (1979)

/14/ Fautz, E., Reichenbach, H.: FEMS Microbiol. Lett. 8, 87-91 (1980)

/15/ Fox, D.L., Lewin, R.A.: Can. J. Microbiol. 9, 753-768 (1963)

/16/ Gripenberg, J.: Acta Chem. Scand. 6, 580-586 (1952)

/17/ Irschik, H., Reichenbach, H.: Biochim. Biophys. Acta 510, 1-10 (1978)

/18/ Jefferson, W.E.: Biochem. 6, 3479-3484 (1967)

/19/ Kanda, N., Ishizaki, N., Inoue, N., Oshima, M., Handa, A., Kitahara, T.: J. Antibiot. 28, 935-942 (1975)

/20/ Lewin, R.A., Lounsbery, D.M.: J. gen. Microbiol. 58, 145-170 (1969)

/21/ McMeekin, T.A., Patterson, J.T., Murray, J.G.: J. appl. Bact. 34, 699-716 (1971)

/22/ Pattenden, G.: Tetrahedr. Lett. 1969, 4049-4052

/23/ Reichenbach, H., Kleinig, H.: Zbl. Bakt., I. Abt. Orig. A 220, 458-463 (1972)

/24/ Reichenbach, H., Kleinig, H., Achenbach, H.: Arch. Microbiol. 101, 131-144 (1974)

/25/ Reichenbach, H., Kohl, W., Böttger-Vetter, A., Achenbach, H.: Arch. Microbiol. 126, 291-293 (1980)

/26/ Simon, G.D., White, D.: Arch. Mikrobiol. 78, 1-16 (1971)

/27/ Sleeper, H.L., Fenical, W.: J. Am. Chem. Soc. 99, 2367-2368 (1977)

/28/ Stodola, F.H., Weisleder, D., Vesonder, R.F.: Phytochem. 12, 1797-1798 (1973)

/29/ Wachter, W.: Eine chemosystematische Studie: Vergleichende Pigmentmusteranalyse der Gattungen Flexibacter, Cytophaga und Flavobacterium. Diplomarbeit, Univ. Freiburg, FRG, 1974, 78 pp.

# PRELIMINARY STUDIES OF THE PIGMENTS OF *FLAVOBACTERIUM BREVE* NCTC 11099 AND *FLAVOBACTERIUM ODORATUM* NCTC 11036

O.B. Weeks

Arts and Sciences Research Center
New Mexico State University
Las Cruces, New Mexico 88003, USA

## Summary

The pigments of *Flavobacterium breve* NCTC 11099 and *Flavobacterium odoratum* NCTC 11036 are extractable with acetone and each consists of a major constituent and three or more minor pigments. The electronic absorption spectrum of the major pigment of each species shows a single maximum absorption and lack of fine structure. The principal pigment of *F. breve* shows maximum absorption at 435 nm in diethyl ether or 442 nm in ethanol. That of *F. odoratum* occurs at 437 nm in diethyl ether or 450 nm in ethanol. Each pigment demonstrates an alkaline-catalyzed bathochromic spectral shift which is completely reversed by acid. *F. breve* shows a bathochromic shift of 36 nm and *F. odoratum*, 28 nm. The pigments are not carotenoid in nature.

## Introduction

*Flavobacterium breve* and *F. odoratum* recently have been subjected to extensive studies which resulted in revised descriptions for the species and the selection of neotype strains [4,5]. *F. breve* has also been proposed as the type species of the genus to replace *F. aquatile* [6]. Chemical studies of the pigments of the two species have been commenced using the neotype designates *F. breve* NCTC 11099 and *F. odoratum* NCTC 11036.

## Results

Preliminary work to select a growth medium for cultivation of F. breve and F. odoratum showed that a medium containing proteose peptone (0.5%), beef extract (0.2%), yeast extract (0.1%), all Difco products, dissolved in 0.01M phosphate buffer, pH 7.0, was suitable and superior to media containing glucose. The culturing system was 1ℓ medium contained in a 2ℓ flask, and incubated at 30°C, 48 hr., with continuous shaking, 250 rpm. The total culture volume used was 8ℓ. Each culture flask was inoculated with 100 ml of a 24 hr.-old primary culture grown in the same medium.

The F. breve culture grew readily, growth after 48 hr. being equivalent to 1.2 OD units at 580 nm (Spectronic 20, Bausch and Lomb). F. odoratum was more difficult to culture seldom producing growth equivalent to 1.0 OD unit. On one occasion the culture lysed after 30 hr. The two bacterial strains have quite different morphologies when grown as in this study (Figs. 1,2).

The bacterial cells were easily removed from the cultures by centrifugation (Sharples Super Centrifuge, Pennsalt Chemicals Co.). The cells were placed in acetone and then stirred for 1 hr., room temperature. The pigment apparently was completely removed by this treatment. The amount of pigment was estimated spectrally using an $E^{1\%}_{cm}$ value of 2000. On this basis F. breve yielded 0.9 mg total pigment from the 8ℓ and this is equivalent to 0.09 mg/OD unit·ℓ. F. odoratum yielded 3.75 mg from the 8ℓ which was equivalent to 0.47 mg/OD unit·ℓ. F. odoratum produced almost 6 times more pigment than F. breve.

The pigment extracts were subjected to phase separation and successive column chromatographic treatments to separate and partially purify the constituents. The acetone was removed (Büchi Rotovapor) and the residue dissolved in chloroform. This solvent was filtered through activated alumina (neutral alumina AG-7, Bio-rad Laboratories) before use to remove the stabilizing ethanol. The chloroform solutions were converted into Bligh and Dyer systems by the addition of methanol and water [3]. After shaking each system was allowed to stand overnight for complete separation of the two phases. The pigment partitioned into the chloroform phase.

The chloroform solutions were partially purified using a silicic acid system [7]. In each instance the entire pigment fraction was eluted with chloroform. The second column system was alumina activity III developed with hexane-acetone mixtures. The chromatographic behavior of the pigments from the two bacterial species was nearly identical. In both instances five colored fractions separated and were eluted from the column. The principal fraction for each species, comprising about 90% of the total pigment separated and eluted with 12% acetone in hexane. Two minor fractions eluted prior to this and another, dark pink, with 15% acetone. Each column chromatographic fraction was chromatographed on Silica Gel G thin-layer plates, 0.25 mm thickness developed with either chloroform: methanol (9:1) or benzene-ethyl ace-

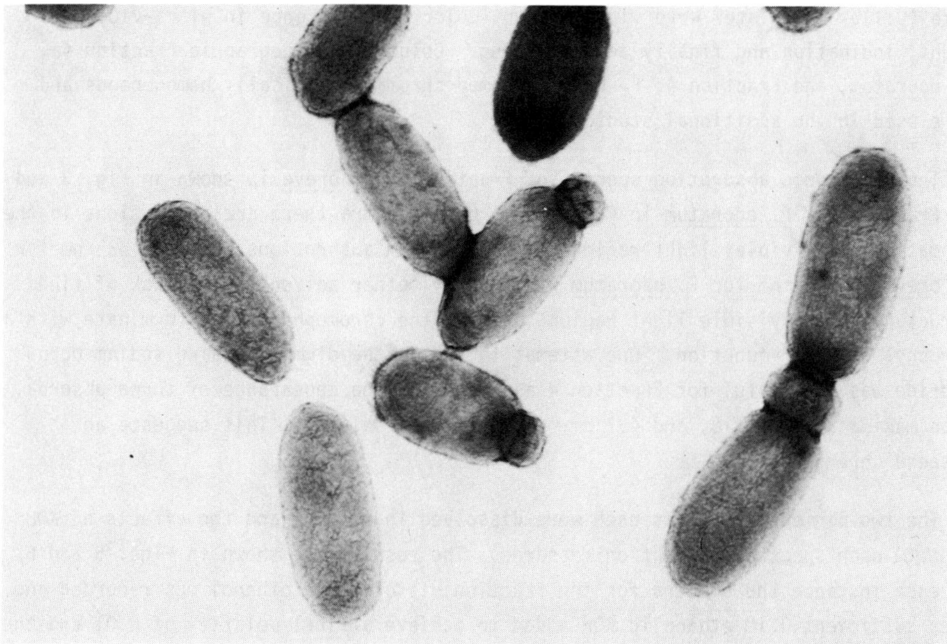

Fig. 1. Electron micrograph of Flavobacterium breve NCTC 11099 stained with phosphotungstic acid, 20,000X magnification.

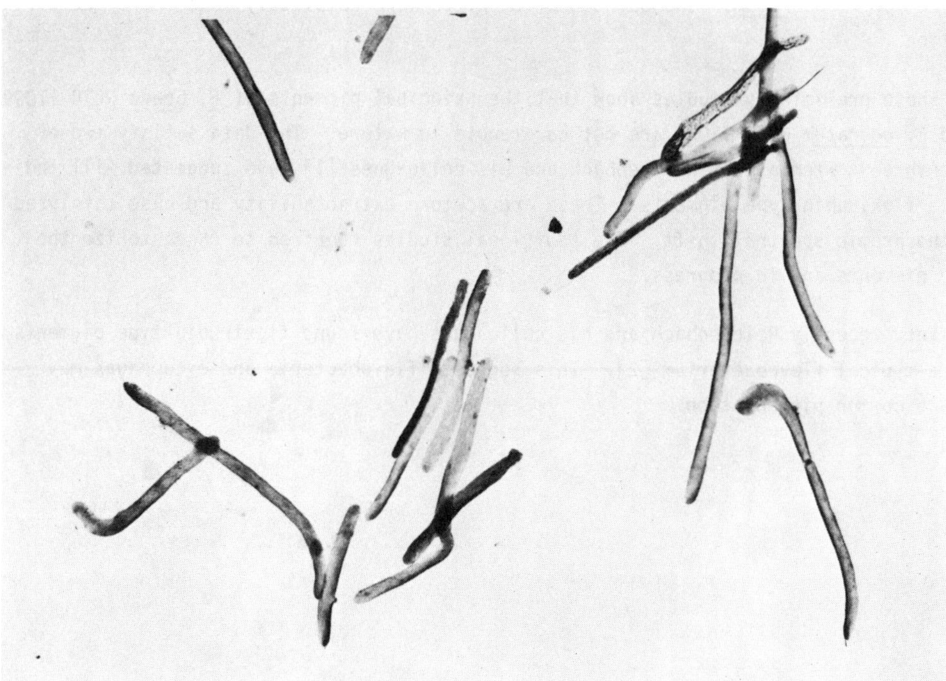

Fig. 2. Electron micrograph of Flavobacterium odoratum NCTC 11036, 42,000 X magnification

tate (9:1). The plates were visualized by color, fluorescence in ultra-violet light, iodination and finally acid charring. Column chromatographic Fraction 4a, F. odoratum, and Fraction 4, F. breve, seemed chromatographically homogeneous and were used in the additional studies.

The electronic absorption spectra of Fraction 4, F. breve is shown in Fig. 3 and of Fraction 4a, F. odoratum in Fig. 4. In both spectra there are absorptions in the aromatic, ultra-violet light region and the maximum absorptions occur at 435 nm for F. breve and 437 nm for F. odoratum with diethyl ether solvent. The lack of fine structure in the visible light region suggests the chromophores may terminate with a carbonyl or ester function. One attempt to reduce the pigments using sodium borohydride was successful for Fraction 4 as judged by the appearance of three absorption maxima at 395, 418, and 437 nm, diethyl ether solvent. This suggests an octaene chromophore.

The two pigment fractions each were dissolved in ethanol and the effects of KOH and HCl upon spectral absorption measured. The results are shown in Figs. 5 and 6. In each instance the spectra for the fraction dissolved in ethanol was recorded and then sufficient 0.1M ethanolic KOH added to achieve a final molarity of 0.01 and the spectra again recorded. Finally the alkaline solutions were acidified using 0.5M ethanolic HCl and the spectra measured. The results show that the absorption maximum for each pigment fraction was bathochromically displaced, 36 nm for F. breve and 28 nm for F. odoratum. The bathochromic shifts were completely reversed by acidification.

These preliminary studies show that the principal pigments of F. breve NCTC 11099 and F. odoratum NCTC 11036 are not carotenoid in nature. The data satisfy two of the three criteria that Reichenbach and his colleagues [1] have suggested will indicate flexirubin-type pigments. These are acetone extractability and base catalyzed bathochromic spectral shift. The additional studies required to characterize the two pigments are in progress.

Very recently Reichenbach and his colleagues have found flexirubin-type pigments in a typical flavobacterium [2]. This suggests flavobacteria and cytophagas may share common pigmentation.

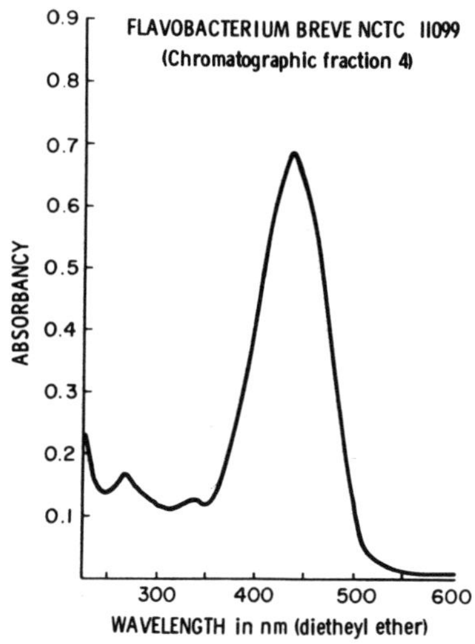


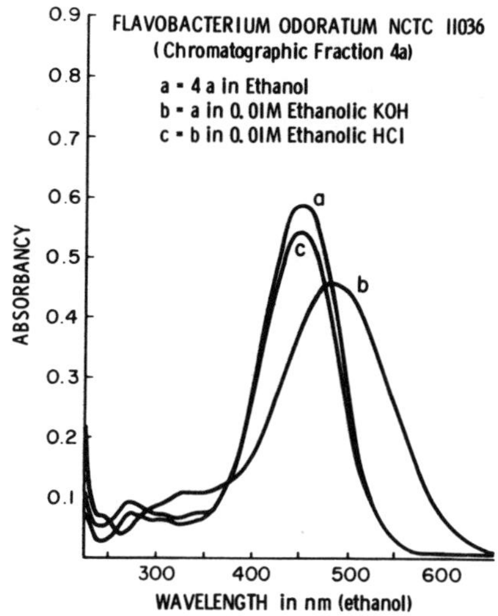


Fig. 3. Electronic absorption spectrum of chromatographic Fraction 4 from Flavobacterium breve NCTC 11099, diethyl ether solvent.

Fig. 4. Electronic absorption spectrum of chromatographic Fraction 4a from Flavobacterium odoratum NCTC 11036, diethyl ether solvent.

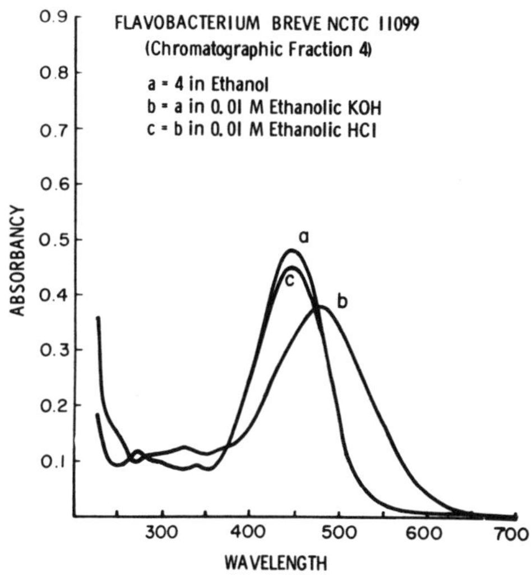


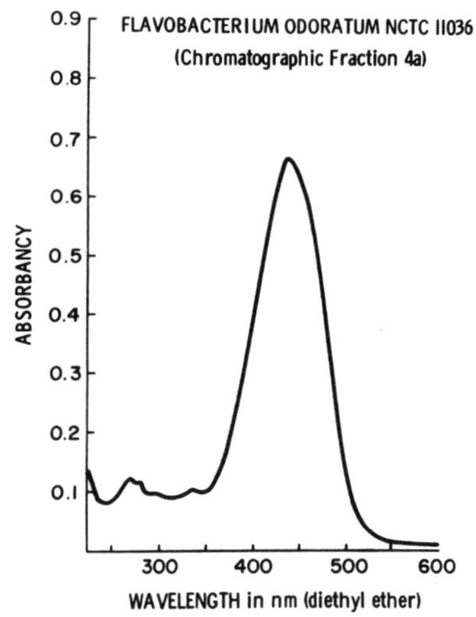


Fig. 5. Electronic absorption spectra of chromatographic Fraction 4 from Flavobacterium breve NCTC 11099 in ethanol (a), 0.01M ethanolic KOH (b) and 0.01M HCl (c).

Fig. 6. Electronic absorption spectra of chromatographic Fraction 4a from Flavobacterium odoratum NCTC 11036 in ethanol (a), 0.01M ethanolic KOH (b) and 0.01 ethanolic HCl (c).

## References


[1] Achenbach, H., Kohl, W., Reichenbach, H.: Revista Latinoamericana de Quimica 9: 111-124 (1978)

[2] Achenbach, H., Kohl, W., Böttger-Vetter, A., Reichenbach, H.: Tetrahedron 1981, in print

[3] Bligh, E.G., Dyer, W.J.: Can. J. Biochem. Physiol. 37, 911-917 (1959)

[4] Holmes, B., Snell, J.J.S., Lapage, S.P.: Int. J. Syst. Bact. 27, 330-336 (1977)

[5] Holmes, B., Snell, J.J.S., Lapage, S.P.: Int. J. Syst. Bact. 28, 201-208 (1978)

[6] Holmes, B., Owen, R.J.: Int. J. Syst. Bact. 29, 416-426 (1979)

[7] Rouser, G., Kritchevsky, G., Yamamoto, A.: p. 99-162 in: Marinetti, G.V. (ed.), Lipid chromatographic analysis. New York (1967): Marcel Dekker.

# TAXONOMICALLY USEFUL TEST PROCEDURES PERTAINING TO BACTERIAL LIPOQUINONES AND ASSOCIATED FUNCTIONS, WITH SPECIAL REFERENCE TO FLAVOBACTERIUM AND CYTOPHAGA

W. Mannheim

Zentrum für Hygiene und Medizinische Mikrobiologie der Philipps-Universität, Abteilung Bakteriologie, Pilgrimstein 2, D-3550 Marburg, F.R.G.

## Summary

The following phenomena related to the respiratory quinones of chemo-organotrophic bacteria are considered regarding their use by the microbiologist: (i) Types and concentrations of quinones present in the cells, (ii) reaction sites of quinones in the respiring membrane, (iii) changes in quinone contents with culture conditions, (iv) effect of fumarate on oxygen-limited growth, (v) electron transport rates in membrane particles with varied donors and acceptors, and (vi) quinone isoprenologs, and degree of unsaturation and branching of the terpenoid side chains. The taxonomic and diagnostic significance of such information is discussed for selected organisms, among them, collection strains representing some *Flavobacterium* and *Cytophaga* species.

## Introduction

Lipoquinones are essential components of membrane-bound electron transport systems not only in mitochondria and chloroplasts but also in many free-living bacteria. Although the biochemistry of the quinones and associated functions are now widely understood, their taxonomic aspects are just beginning to be realized by microbiologists. Some of the methods required can be easily adapted to the microbiological laboratory. In the following, several phenomena related to the respiratory quinones of chemo-organotrophic bacteria shall be stressed, emphasizing the use of such criteria as guides in the classification and identification, particularly of bacteria of the *Flavobacterium*-*Cytophaga* complex.

## Quinone Contents

Respiratory quinones are lacking in some groups of anaerobic or aerotolerant organisms which are unable to conserve energy by oxidative phosphorylation coupled to membrane-associated electron transport, i.e. clostridia and mycoplasmas, lactobacilli and streptococci - with the unique exception of S. faecalis which contains demethylmenaquinones /2/ - and peptococci /cf. 22, 27/. On the other hand, the presence of respiratory capabilities is strongly correlated with the presence of quinones in the micromolar order of magnitude per gram of cell protein which are located in the lipid phase of the cytoplasmic membrane. (Lipoquinones present in the nanomolar order of magnitude are not involved in membrane-associated electron transport /27/ and shall not be dealt with here.) The distribution of respiratory quinones among the chemo-organotrophic bacteria suggests the general view that larger, i.e. generic or supra-generic, groups of bacteria are homogeneous with respect to the types of quinones synthesized. Menaquinones are produced by the Gram-positive, aerobic and facultative groups throughout (e.g., staphylococci and micrococci /31, 60/, bacilli /4, 17, 22, 37, 54/, actinomycetes and related organisms /1, 3, 4, 10, 11, 45, 61/), by Gram-negative, anaerobic bacteria (e.g. Bacteroides and related groups /12, 20, 21/ and Desulfovibrio /44, 47, 55/) and furthermore, by some Gram-negative, aerobic, and mostly gliding bacteria (Cytophaga /7/, Flexibacter /33/, some myxobacteria /33/ and flavobacteria /7/). Ubiquinones are the typical respiratory quinones of the strictly aerobic, Gram-negative rods and cocci /4, 15, 26, 42, 43, 50, 59, cf. 14/ and among them, some flavobacteria /7, 15/; ubiquinones are also the sole respiratory quinones in some groups of fermentative, Gram-negative organisms (e.g. Cardiobacterium /42/, Erwinia /25/ and Kingella /42/). The majority of the Gram-negative, facultative, and fermentative bacteria are capable of producing both ubiquinones and menaquinones and/or demethylmenaquinones (i.e., Enterobacteriaceae and Vibrionaceae /4, 30, 38, 39, 49, 56, 57/ and Pasteurellaceae /4, 23, 26, 26, 40, 42/ with the exception of some groups of Pasteurellaceae that contain only demethylmenaquinones /26, 40, 42/).

The type of quinone used in a respiratory chain system provides fundamental physiotaxonomic information . According to the different redox potentials of the quinones, ubiquinone can only be used in electron transport to high-potential acceptors, such as oxygen and nitrates, whereas the naphthoquinones are also capable of mediating anaerobic electron transport to low-potential acceptors, such as, fumarate, provided the appropriate enzymes be available /24, 37/. Organisms containing both benzo- and naphthoquinones display a high metabolic versatility by the alternative use of aerobic and anaerobic electron transport systems according to the terminal acceptor available. Providencia rettgeri /38/ and Haemophilus species /23, 24/ are typical examples of the branched type of electron transport systems. Considering their physiologic significance and distribution the respiratory quinones undoubtedly should be included in lists of minimal standards for the characterization of bacterial taxa.

The procedures for determination of the quinone contents are easy to perform. The quinones can be isolated by solvent extraction from wet materials, and samples containing 5 to 10 mg of bacterial protein are usually sufficient. Sonic homogenates of cells suspended in neutral tris-hydrochloride (not phosphate) buffer are oxidized with hydrogen peroxide (plus catalase, where necessary), extracted with a mixture of methanol-light petroleum and acetone, according to Kröger /36, 37, 38/, and identified by either spectrophotometry or thin-layer chromatography, or both. Thin-layer chromatography that needs no special equipment may be used for the preliminary identification, and the preparative separation of mixtures of benzoquinones and naphtoquinones. The dry residue of the light petroleum extract is redissolved in a few drops of light petroleum, and chromatographed on kieselgel 60 (nr. 5721, E. Merck, Darmstadt, Germany) with benzene-chloroform 9:1 (v/v) as solvent for about 30 min., and the spots are traced with the bispyridyl-ferrichloride reagent /56/ after reduction with a potassium borohydride spray and neutralization with a hydrochloric acid spray. In this system, the Rf values of the ubiquinones range between 0.10 and 0.25, and those of the naphthoquinones between 0.4 and 0.6 with a limit of detection of about 10 nmol of either quinone (cf. /27/, for other systems, see /5/). The thin-layer chromatogram does not discriminate between menaquinones and demethylmenaquinones. The quantitative determination and identification of the quinones present in the crude lipid extract, or after purification by thin-layer chromatography is achieved by the ultra-violet difference spectrum of the reduced minus oxidized state in acetate-buffered ethanolic solution, using an appropriate double-beam recording spectrophotometer (for ubiquinones and menaquinones, see /36, 37, 38/, for demethylmenaquinones, see /2, 40/).

The quinones present in some representative Flavobacterium and Cytophaga strains are shown by Table 1. All strains containing ubiquinones are labelled Flavobacterium, among them, the type strain of the type species, F. aquatile /29/. The menaquinone-forming strains represent either Flavobacterium, or Cytophaga species. One of them, F. heparinum has only recently been transferred to Cytophaga /9/ on the basis of independent arguments. If a ubiquinone-producing organism remained type species of the genus Flavobacterium, the menaquinone-forming strains would have to be eliminated since one bacterial genus cannot comprise organisms that use exclusively ubiquinones and organisms using only naphthoquinones in their respiratory chain system. There is some evidence from DNA:DNA hybridization experiments that several menaquinone-producing flavobacteria with low G+C contents are phylogenetically related - though with low DNA binding levels - to members of the family Cytophagaceae /8/, a bacterial group in which menaquinone appears to be used as sole respiratory quinone throughout /7, 32, 33/. Following this view, the family Cytophagaceae should be considerably enlarged, by the "low-GC" members of the Flavobacterium-Cytophaga complex producing menaquinones, and the genus Flavobacterium confined to the "low-GC" organisms containing ubiquinones.

Table 1. Respiratory Quinones of Some Flavobacterium and/or Cytophaga Strains[1]

| G+C moles% | Designation of organism | Collection number | Quinone[2] |
|---|---|---|---|
| 32 | F. aquatile (T) | ATCC 11947 | Q |
| 50 | F. halmephilum (T) | ATCC 19717 | Q |
| 63 | F. capsulatum (T) | ATCC 14666 | Q |
| 68 | Flavobacterium sp. | ATCC 19945 | Q |
| 69 | F. devorans (T) | ATCC 10829 | Q |
| 26/34 | F. breve | ATCC 14234 | MK |
| 32 | F. uliginosum (T) | ATCC 14397 | MK |
| 33/35 | F. pectinovorum | ATCC 19366 | MK |
| 33/35 | C. johnsonae (T) | ATCC 17061 | MK |
| 34 | F. tirrhenicum | ATCC 15997 | MK |
| 36 | Flavobacterium/Cytophaga sp. | NCMB 251 | MK |
| 37 | F. odoratum (T) | ATCC 4651 | MK |
| 36/38 | F. meningosepticum serogr.A (T) and serogroups B - F | NCTC 10016, NCTC 10585-10589 | MK |
| 36/39 | Flavobacterium sp., group IIb | NCTC 10795-10797 | MK |
| 37 | C. marinoflava | ATCC 19326 | MK |
| 39 | C. hutchinsonii (T) | NCIB 9469 | MK |
| 41 | Flavobacterium/Cytophaga sp. | NCMB 249 | MK |
| 42 | C. fermentans (T) | ATCC 19072 | MK |
| 42 | C. aurantiaca | ATCC 12208 | MK |
| 42/46 | F. heparinum (T) | ATCC 13125 | MK |
| 43 | F. ferrugineum (T) | ATCC 13524 | MK |
| 63/64 | Flavobacterium/Cytophaga sp. | NCMB 244 | MK |

1/ Data from Callies and Mannheim,/7/
2/ Quinones: MK = menaquinones, Q = ubiquinones
T = Recognized Type strain according to Approved Lists of Bacterial Names /29/

If however, F. breve which contains menaquinone is to be recognized as the type species as has been proposed /28/, the genus Flavobacterium will probably have to be transferred to the family Cytophagaceae, and the "low-GC" ubiquinone-forming strains will become taxonomic orphans, as do the strains with high G+C contents.

## Reaction Sites of Quinones in the Respiring Membrane

The quinone extraction-reincorporation technique /18, 23, 24, 37, 38/ yields valuable additional information on quinone reaction sites in the respiring membrane. The procedure, however, is laborious and time-consuming, and cannot therefore be applied on a large scale. Nevertheless, the data available appear to represent the general behavior of certain types of bacterial membranes with respect to the quinone reaction sites. Membranes of aerobic Gram-negative bacteria which only produce ubiquinones are incapable of utilizing naphthoquinones in their respiratory chain system (unpublished data of R. Holländer for Pseudomonas fluorescens and Acinetobacter sp.) and inversely, Gram-positive bacteria producing only naphthoquinones do not display the reaction sites for ubiquinones /37/. Flavobacterium meningosepticum strain King 14 behaves like the Gram-positives in this respect /6/, and this is probably the case with the other menaquinone-containing Flavobacterium and Cytophaga strains. Menaquinones and demethylmenaquinones can replace each other in naphthoquinone-mediated electron transport pathways except for succinate oxidation which is supported by demethylmenaquinone but not by menaquinone /24/. With bacterial groups that use both ubiquinones and naphthoquinones, the respiring membrane exhibits two different quinone reaction sites, one for each type of quinone /38/. These sites are maintained even in strains with a recent genetic block of quinone biosynthesis. E.g., ubiquinones restore the electron transport to oxygen in quinone-depleted membranes of Haemophilus influenzae, a member of the Pasteurellaceae that actually produces only demethylmenaquinones /23/. Similarly, the conservation of the reaction site for naphthoquinones in a mutant strain of Haemophilus parainfluenzae (Pittman 429) lacking naphthoquinone production has been observed /23/. Genetic variation in bacterial quinone biosynthesis may be obtained under experimental conditions /13, 48, 53, and others/ but appears rarely to occur in the natural environment, except perhaps with obligate parasites.

## Changes in Quinone Contents with Culture Conditions

The phenotypic expression of respiratory chain components is highly dependent on the nature and concentration of the terminal electron acceptor available during growth. E.g. the ubiquinone contents of an Acinetobacter strain increased about eight to tenfold between oxygen-saturation and the oxygen limited state of the culture /19/. Similar data are available for the ubiquinone contents of Paracoccus denitrificans with oxygen and nitrate as terminal acceptors /43/, and for the dynamics of the respiratory chain of a demethylmenaquinone-containing Haemophilus parainfluenzae strain /58/. With organisms producing both ubiquinones and naphthoquinones, the former predominate in the aerobic culture whereas the latter prevail in anaerobic conditions provided a suitable terminal acceptor for anaerobic electron transport be present. Although the quinone biosynthesis may be partly constitutive, the full range of variation in quinone contents with such organisms is only detected by comparative analysis of cells grown with alternative electron acceptors, e.g. aerobically, and anaerobically with fumarate /26, 30, 38, 42, 56, 57/.

Besides the nature of the terminal electron acceptor available, the composition of the culture medium may influence the biosynthesis of quinones. E.g. in Bacteroides fragilis the biosynthesis of menaquinone appears to be repressed by glucose. In some Pasteurellaceae that produce only demethylmenaquinone in carbohydrate-free peptone media, small amounts of ubiquinones appear in the aerobic culture after addition of glucose (Reihse-Hess and Mannheim, unpublished data). The use of exogeneous naphtoquinones as growth factors by several Bacteroides species /20, 21, 41/ are further examples of possible medium effects on quinone contents.

## Effect of Fumarate on Oxygen-Limited Growth

In many instances, the ability to use fumarate as an alternative electron acceptor can be detected by just comparing the relative turbidities of oxygen-limited cultures with and without added fumarate. Stimulation of the oxygen-limited growth by fumarate in our experience indicates reliably the presence of a naphthoquinone-mediated fumarate respiration system. This simple test procedure can be used in the diagnostic routine laboratory for all organisms that grow in stable suspension. In general we use a proteose peptone medium (Proteose peptone nr. 3, Difco Laboratories, 2% w/v; NaCl 0.5% w/v; tris (hydroxymethyl) aminomethane q.s., pH 7.2) with 0.02 M disodium fumarate, sterilized by filtration, or 0.01 M sodium hydrogencarbonate as supplements for test and control cultures. The media are distributed in 10 ml amounts into 13 mm wide test tubes, inoculated in parallel with small inocula, and the turbidities are compared by visual reading, or photometrically at 600 nm, after 12, 24, or more hours of static incubation at the appropriate temperature. With facultative-aerobic organisms, anaerobic conditions develop during the initial phase of growth, as with F. meningosepticum (Fig. 1) whereas for strictly anaerobic organisms, deoxygenated media and a reducing atmosphere are required.

Among the members of the Flavobacterium-Cytophaga complex, only F. meningosepticum, the group IIb organisms, and F. odoratum exhibited significantly increased growth yields in this test that indicates an energetically useful fumarate respiration /7/. Elsewhere, the test is generally positive with fermentative Gram-negative organisms containing naphthoquinones (with some exceptions) and with several groups of Gram-positive bacteria. However, with staphylococci and the Bacillus sp., which are known to possess the enzymes of fumarate respiration /34, 37, and unpublished data/ the increase in turbidity with fumarate is not so pronounced.

## NADH-Fumarate Reductase Activity

Photometric recording of NADH oxidation by cell homogenates or membrane particles with fumarate as the terminal electron acceptor is a more sensitive means to detect

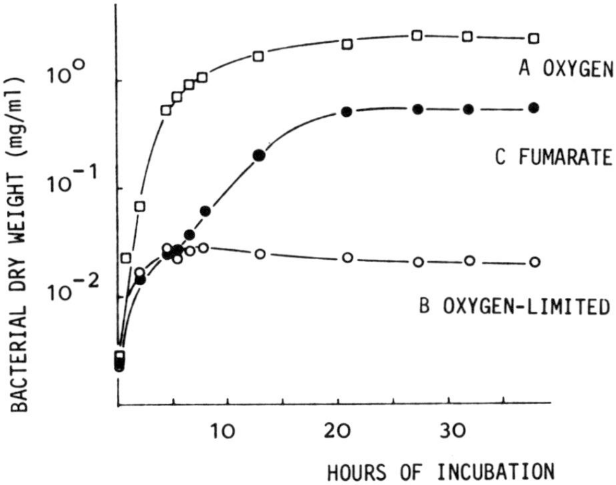


Fig. 1 Growth of Flavobacterium meningosepticum King 14 in Proteose peptone medium A) tubes containing 5 ml of medium, moderately aerated by shaking, B) tubes containing 10 ml of medium and incubated statically, and C) same as (B), but supplemented with 50 mM disodium fumarate. Media were oxygen-saturated at the time of inoculation. From E. Callies, Thesis, Marburg 1979.

naphthoquinone-mediated fumarate respiration. Washed cells suspended in neutral phosphate buffer are sonicated to uncouple phosphorylation from electron transport, and to destroy permeability barriers for NADH. The homogenate is supplemented with excess NADH and the decrease in extinction at 360 nm is recorded until the reaction becomes oxygen-limited; the "NADH oxidase" activity is expressed as micromoles of NADH consumed per minute and gram of bacterial protein. Then fumarate is added to the reaction mixture and after consumption of traces of contaminating oxygen by the NADH oxidase reaction, the "NADH-fumarate reductase" activity is recorded. The NADH-fumarate reductase activities are high in cells grown anaerobically with fumarate, and low in aerobic cultures. In the absence of fumarate, the asparate content of peptone media may be responsible for the induction of the fumarate respiration system. Not all organisms producing naphthoquinones, however, have retained the fumarate respiration pathway.

Membrane-bound fumarate-reductase systems are constantly naphthoquinone-dependent /cf. 35/. On the other hand, soluble NADH-fumarate reductase systems which are free of naphthoquinones and not connected with fumarate respiration do occur, e.g. in most strains of Erwinia herbicola, an organism containing only ubiquinone /25/. To discriminate the membrane-bound and soluble NADH-fumarate reductases, high-speed centrifugation, and comparison of the activities of supernatant and washed pellet fractions of the cell homogenates is necessary.

Most of the fermentative, naphthoquinone-containing Gram-negative and Gram-positive bacteria display membrane-bound NADH-fumarate respiration systems, as do some of the menaquinone-producing Flavobacterium and Cytophaga strains /7/. The NADH-fumarate reductase activities are usually high in organisms exhibiting a strong reaction to fumarate in the growth test, and vice versa. High NADH-fumarate reductase activities without a marked growth response to fumarate suggest the presence of the soluble system. The analysis of the respiratory activities can be extended to other electron donors, such as formate, hydrogen, etc., or other acceptors, like nitrates, sulfates, etc. /cf. 34/.

## Quinone Isoprenologs, Degree of Unsaturation and Branching of the Terpenoid Side Chain

Whereas the distribution of respiratory benzo- and naphthoquinones characterizes large groups of bacteria, the length, and further modifications of the terpenoid side chain are characteristic of smaller taxonomic entities, i.e. species or subspecies. Such data are available for some Micrococcaceae /4, 31, 52, 60/, various groups of Gram-positive rods /1, 3, 4, 10, 11, 16, 17, 22, 45, 54, 60, 61, 62, 64, and others/, Streptococcus faecalis /2/, some anaerobic, Gram-negative bacteria /12, 44, 47, 55/, various strains of Enterobacteriaceae and Vibrionaceae /2, 4, 15, 39, 49, 56, 57,and others/ and Pasteurellaceae /40, Hess-Reihse, unpublished data/, Acetobacter and Gluconobacter /59, 63/, other strictly aerobic Gram-negative bacteria /4, 14, 15, 56, and others/, and some Cytophagaceae /32, 33/. Quinone-isoprenologs of single strains of ubiquinone-forming /15/ and menaquinone-forming /17/ Flavobacterium strains have been reported but so far, a systematic study of the Flavobacterium-Cytophaga complex is lacking. Undoubtedly, the patterns of the quinone isoprenologs would provide valuable additional criteria for the fine classification of those organisms.

The methods include preparative thin-layer chromatography as mentioned above, and analytical reversed-phase thin-layer chromatography to separate the isoprenologs, e.g. on kieselgel 60 HPTLC plates (without indicator, nr. 5631 from Merck, Darmstadt), impregnated with paraffin oil and with acetone-water 95:5 (v/v) as solvent /10, 17, 27/. Argentation may be necessary to detect unsaturation of the side chains /17/ which frequently occurs in the menaquinones of Gram-positive rod-shaped bacteria /1, 3, 10, 11, 17, 61/ and for the final confirmation of the findings, NMR and mass spectroscopy /cf. 51, 61/.

References

/1/ Azerad, R., Cyrot-Pelletier, M.O.: Biochimie 55, 591-603 (1973)

/2/ Baum, R.H., Dolin, M.: J. Biol. Chem. 240, 3425-3433 (1965)

/3/ Beau, S., Azerad, R., Lederer, E.: Bull. Soc. Chim. Biol. 48, 569-581 (1966)

/4/ Bishop, D.H.L., Pandya, K.P., King, H.K.: Biochem. J. 83, 606-614 (1962)

/5/ Bolliger, H.R., König, A.: p. 253-302 in: Stahl, E. (ed.), Dünnschicht-Chromatographie, 2nd. ed. Berlin, Heidelberg, New York (1967): Springer

/6/ Callies, E.: Thesis, Fachbereich Biologie der Philipps-Universität Marburg (1979)

/7/ Callies, E., Mannheim, W.: Int. J. Syst. Bacteriol. 28, 14-19 (1978)

/8/ Callies, E., Mannheim, W.: Antonie van Leeuwenhoek 46, 41-49 (1980)

/9/ Christensen, P.: Int. J. Syst. Bacteriol. 30, 473-475 (1980)

/10/ Collins, M.D., Jones, D., Goodfellow, M., Minnikin, D.E.: J. Gen. Microbiol. 111, 453-457 (1979)

/11/ Collins, M.D., Piroutz, T., Goodfellow, M., Minnikin, D.: J. Gen. Microbiol. 100, 221-230 (1977)

/12/ Collins, M.D., Shah, N.H., Hardie, J.M.: Abstract nr. 10 in: Czechoslovak Society for Microbiology, Conference on Taxonomy, Computer Identification of Bacteria and Diagnostik Methods, held in Liblice Castle, 5.-9. May 1980

/13/ Cox, G.B., Gibson, F., Pittard, J.: J. Bacteriol. 95, 1591-1598 (1968)

/14/ Crane, F.L.: p. 183-206 in Morton, R.A. (ed.), Biochemistry of Quinones. London, New York (1965): Academic Press

/15/ Denis, F.A., D'Oultremont, P.A., Debacq, J.J., Cherel, J.M., Brisou, J.: C.R. Soc. Biol. 169, 380-383 (1975)

/16/ Dunphy, P., Brodie, A.F.: Methods Enzymol. 18C, 407-461 (1971)

/17/ Dunphy, P.J., Phillips, P.G., Brodie, A.F.: J. Lipd. Res. 12, 442-449 (1971)

/18/ Ernster, L., Lee, I.-Y., Norling, B., Person, B.: Europ. J. Biochem. 9, 299-310 (1969)

/19/ Fischer, P.: Thesis, Fachbereich Humanmedizin der Philipps-Universität Marburg (1972)

/20/ Gibbons, R.J., Engle, L.P.: Science 146, 1307-1309 (1964)

/21/ Gibbons, R.J., MacDonald, J.B.: J. Bacteriol. 80, 164-170 (1960)

/22/ Hess, A., Holländer, R., Mannheim, W.: J. Gen. Microbiol. 115, 247-252 (1979)

/23/ Holländer, R.: Thesis, Fachbereich Biologie der Philipps-Universität Marburg (1976)

/24/ Holländer, R.: FEBS Letters 72, 98-100 (1976)

/25/ Holländer, R.: Zbl. Bakt. Hyg. I. Orig. C 1 (1980 in the press)

/26/ Holländer, R., Mannheim, W.: Int. Syst. Bacteriol. 25, 102-107 (1975)

/27/ Holländer, R., Wolf, G., Mannheim, W.: Antonie van Leeuwenhoek 43, 177-185 (1977)

/28/ Holmes, B., Owen, R.J.: Int. J. Syst. Bacteriol. 29, 416-426 (1979)

/29/ International Committee on Systematic Bacteriology, Judicial Commission (Skerman, V.D.B., McGowan, V., Sneath, P.H.A. eds.): Int. J. Syst. Bacteriol. 30, 225-420 (1980)

/30/ Jedif3, R.: Thesis, Fachbereich Humanmedizin der Philipps-Universität Marburg (1973)

/31/ Jeffries, L.: Int. J. Syst, Bacteriol. 19, 183-187 (1969)

/32/ Kleinig, H., Reichenbach, H.: Arch. Microbiol. 112, 307-310 (1977)

/33/ Kleinig, H., Reichenbach, H., Theobald, N., Achenbach, H.: Arch. Microbiol. 101, 91-93 (1974)

/34/ Kröger, A.: p. 61-93, in Microbial Energetics. Symp. Soc. Gen. Microbiol. XXVII. Cambridge (1977): University Press

/35/ Kröger, A.: Biochem. Biophys. Acta 505, 129-145 (1978)

/36/ Kröger, A.: Methods Enzymol. 53D, 579-591 (1978)

/37/ Kröger, A., Dadâk, V.: Europ. J. Biochem. 11, 328-340 (1969)

/38/ Kröger, A., Dadâk, V., Klingenberg, M., Diemer, F.: Europ. J. Biochem. 21, 322-333 (1971)

/39/ Lester, R.L., Crane, F.L.: J. Biol. Chem. 234, 2169-2175 (1959)

/40/ Lester, R.L., White, D.C., Smith, S.L.: Biochemistry 3, 949-954 (1964)

/41/ Lev, M.: J. Gen. Microbiol. 20, 679-703 (1959)

/42/ Mannheim,W., Stieler, W., Wolf, G., Zabel, R.: Int. Syst. Bacteriol. 28, 7-13 (1978)

/43/ Manousek, Z., Safarcīk, K., Dadâk, V.: Scripta Fac. Sci. Nat. UJEP Brunensis, Chemia 2, 6, 61-66 (1976)

/44/ Maroc, J., Azerad, R., Kamen, M.D., Le Gall, J.: Biochem. Biophys. Acta 197, 87-89 (1970)

/45/ Minniki, D.E., Goodfellow, M.: p. 160-219, in: Goodfellow, M., Brownell, G.H., Serrano, J.A.: The Biology of the Nocardiae. London, New York, San Francisco (1976): Academic Press

/46/ Morton, R.A.: p. 1-21, in Morton, R.A.(ed.), The Biochemistry of Quinones. London, New York (1965): Academic Press

/47/ Morton, R.A., Weber, M.M., Matschiner, J.T., Peck, H.D.: Biochem. Biophys. Res. Commun. 38, 197-204 (1970)

/48/ Newton, N.A., Cox, G.B., Gibson, F.: Biochim. Biophys. Acta 244, 155-166 (1971)

/49/ Rebel, G., Mandel, P.: Biochim. Biophys. Acta 98, 380-393 (1965)

/50/ Redfearn, E.: Vitamins & Hormones 24, 465-488 (1966)

/51/ Thomson, R.H.: Naturally Occurring Quinones, 2nd. ed., London, New York (1971): Academic Press

/52/ Threllfall, D.R.: Vitamins & Hormones 21, 153-200 (1971)

/53/ Wallace, B.J., Young, I.G.: Biochim Biophys. Acta 461, 75-83 (1977)

/54/ Watanuki, M., Aida, K.: J. Gen. Appl. Microbiol. 18, 469-472 (1972)

/55/ Weber, M.M., Matschiner, J.T., Peck, H.D.: Biochem. Biophys. Res. Commun. 38, 197-204 (1970)

/56/ Whistance, G.R., Dillon, J.F., Threllfall, D.R.: Biochem. J. 111, 461-472 (1969)

/57/ Whistance, G.R., Threllfall, D.R.: Biochem. J. 108, 505-507 (1968)

/58/ White, D.C.: J. Bacteriol. 93, 567-573 (1967)

/59/ Yamada, Y., Aida, K., Uemura, T.: J. Gen. Appl. Microbiol. 15, 181-196 (1969)

/60/ Yamada, Y., Inouye, G., Tahara, Y., Kondo, K.: J. Gen. Appl. Microbiol. 22, 203-214; 227-236 (1976)

/61/ Yamada, Y., Inouye, G., Tahara, Y., Kondo, K.: Biochim. Biophys. Acta 486, 195-203 (1977)

/62 Yamada, Y., Ishikawa, T., Tahara, Y., Kondo, K.: J. Gen. Appl. Microbiol. 23, 207-216 (1977)

/63/ Yamada, Y., Nakazawa, E., Nozaki, A., Kondo, K.: J. Gen. Appl. Microbiol. 22, 285-292 (1976)

/64/ Yamada, Y., Yamashita, M., Tahara, Y., Kondo, K.: J. Gen. Appl. Microbiol. 23, 331-335 (1977)

# ADDENDA ET CORRIGENDA

## LIPOQUINONES OF FLAVOBACTERIUM AQUATILE STRAIN TAYLOR F 36

W. Mannheim

Zentrum für Hygiene und Medizinische Mikrobiologie der Philipps-Universität
Abteilung Bakteriologie, Pilgrimstein 2, D-3550 Marburg, F.R.G.

Since discussions during the meeting in Braunschweig-Stöckheim, July 8-11, 1980 have shown that the quinone type of *Flavobacterium aquatile* strain Taylor F 36 as published by Callies /1/ and Callies and Mannheim /2/ does not agree with several other chemotaxonomic systems, four different lines of that organism were re-examined for their quinones: (i) HIM 446-5, derivative of ATCC 11947, (ii) strain of O.B. Weeks (identical with ATCC 11947), kindly provided by K. Kersters, Gent, (iii) NCIB 8694, original, and (iv) NCIB 8694, kindly provided by K. Kersters. The four lines were very similar in their phenotypes and therefore could be considered as authentic Taylor F 36 cultures. Mass cultures grown on the medium of Weeks and Beck /3/ and carefully tested for purity were found to contain menaquinones, in the micromolar order of magnitude per gram of cellular protein.

Thus the statement that *F. aquatile* belongs to the ubiquinone-forming category /4/ must be corrected, and corresponding taxonomic considerations changed accordingly. With this correction, the quinone types reported for representative *Flavobacterium* and *Cytophaga* species are in full accordance with groupings based on the base sequence relatedness of the ribosomal cistrons /5/ or the classification of other cellular lipids /6/ and therefore, corroborate the reliability of the method as a taxonomic guide with the "low G+C" organisms of the *Flavobacterium*-*Cytophaga* complex.

### References

/1/ Callies, E.: Thesis, Fachbereich Biologie der Philipps-Universität Marburg (1979)

/2/ Callies, E., Mannheim, W.: Int. J. Syst. Bacteriol. 28, 14-19 (1978)

/3/ Weeks, O.B., Beck, S.M.: J. Gen. Microbiol. 23, 217-229 (1960)

/4/ Mannheim, W.: This Symposium

/5/ De Ley, J.: This Symposium

/6/ Reichenbach, H.: Personal communication

# HYDROXY FATTY ACIDS AS VALUABLE CHEMOSYSTEMATIC MARKERS IN GLIDING BACTERIA AND FLAVOBACTERIA

E. Fautz[1], L. Grotjahn, H. Reichenbach

Gesellschaft für Biotechnologische Forschung,
Mascheroder Weg 1, D-3300 Braunschweig-Stöckheim, FRG.
[1]Present address: Biologisches Institut II der Universität,
Lehrstuhl für Biochemie der Pflanzen, D-7800 Freiburg, FRG

## Summary

The fatty acid patterns of Cytophaga-like bacteria are dominated by 2- and 3-hydroxy fatty acids with iso-branched carbon-chains. Several flavobacteria with DNA in the low GC-range have patterns nearly identical to those from cytophagas. With other flavobacteria, particularly such with DNA in the high GC-range, fatty acids with branched chains are almost completely lacking, and hydroxy fatty acids are present in small quantities only. Myxobacteria clearly form two groups with respect to their fatty acid patterns. While the representatives of the suborder Cystobacterineae (Myxococcus, Cystobacter, Stigmatella) contain substantial quantities of hydroxy fatty acids, the representatives of the suborder Sorangineae (Sorangium, Nannocystis) contain no hydroxy fatty acids at all, at least not in total cell hydrolysates. The main fatty acids of all myxobacteria have iso-branched carbon chains.

As has been known for some time, the fatty acid patterns of gliding bacteria are usually dominated by compounds with (iso-) branched carbon chains /3, 9, 10, 11, 12, 13/. This is uncommon for strictly aerobic gram-negative bacteria. Recently another peculiarity of the fatty acid patterns of gliding bacteria has been discovered: i.e. that many of these organisms contain substantial amounts of hydroxy fatty acids /3/. In some cases up to 50 % of the total fatty acids were hydroxylated. While up to this time hydroxy fatty acids had only been demonstrated in lipopolysaccharide preparations from gliding bacteria /9/, they have now been found even in total cell hydrolysates, and in addition to the more common 3-hydroxy compounds, 2-hydroxy fatty acids could be identified. Branched-chain 2-hydroxy fatty acids are quite unusual and seem not to have been isolated from living organisms before. It was interesting to see, that the different groups of gliding bacteria each seem to have their own special fatty acid patterns, with the characteristic differences usually in the hydroxy acid fraction. Thus Flexibacter elegans and several other Cytophaga-like bacteria (CLB) have as the main components (besides $C_{16}$, $C_{16:1}$ and iso $C_{15}$) 2- and 3-OH iso $C_{15}$, and 3-OH iso $C_{17}$ acids, while the myxobacteria Myxococcus fulvus and Stigmatella aurantiaca yielded mainly (besides $C_{16}$, $C_{16:1}$ and iso $C_{15}$) iso $C_{17}$ and 2-OH iso $C_{17}$ acids. It thus appeared to us that the fatty acid pattern might be used in this group of bacteria as a systematic lead. We therefore analyzed more strains and species of gliding bacteria for their fatty acid patterns, and also included several flavobacteria in this study, because there may be taxonomic relations between certain flavobacteria and some of the CLB.

## Materials and Methods

Organisms. Flexibacter elegans Fx el, marine Cytophaga-like bacterium (CLB) Sy 126, and most myxobacteria were from our own collection. Sorangium cellulosum ssp. fulvum So ce14 is ATCC strain 25532 (producer of the antibiotic ambruticin). Cytophaga johnsonae Cy j1 is NCIB strain 10150, Cy. aquatilis Cy aq1 was kindly supplied by W.S. Strohl, Baton Rouge, marine CLB Cy. latercula Cy la1 is ATCC strain 23177. Flavobacterium aquatile Fv a1 is ATCC strain 11947, Fv. breve Fv b1 is ATCC strain 14234, Fv. ferrugineum Fv f1 is ATCC strain 13524 (type). Fv. devorans Fv d1 is ATCC strain 10829, Fv. meningosepticum Fv m1 is NCTC strain 10016, Fv. capsulatum Fv c2 is ATCC strain 14666 (type), flavobacteria C 1/2, C 20/1, B 4/2, and L 10/1 were kindly supplied by Dr. P.R. Hayes, Leeds /5, 6/, and yellow pigmented strains Fv 13-78 and Fv b IV by Dr. W. Loeffler, Tübingen /4/.

Analytic methods were as described previously /3/: Hydrolysis of wet bacteria with methanolic HCl, extraction of fatty acids with petroleum ether, preparation of methyl esters, and their analysis by gas-liquid chromatography.

## Results and Discussion

The fatty acid patterns of representative strains are shown in Figures 1 to 3. The patterns of more than 20 strains of Cytophaga-like bacteria (CLB) investigated were rather uniform (Fig. 1). As was to be expected all profiles were dominated by branched-chain compounds. Hydroxylated fatty acids were always present in large quantities, i.e. in proportions between 30 and 55 % of the total fatty acids, with the main representatives 2-OH $C_{15}$ iso, 3-OH $C_{15}$ iso, and particularly 3-OH $C_{17}$ iso (in marine Cytophaga latercula the latter compound reached 42 %!).

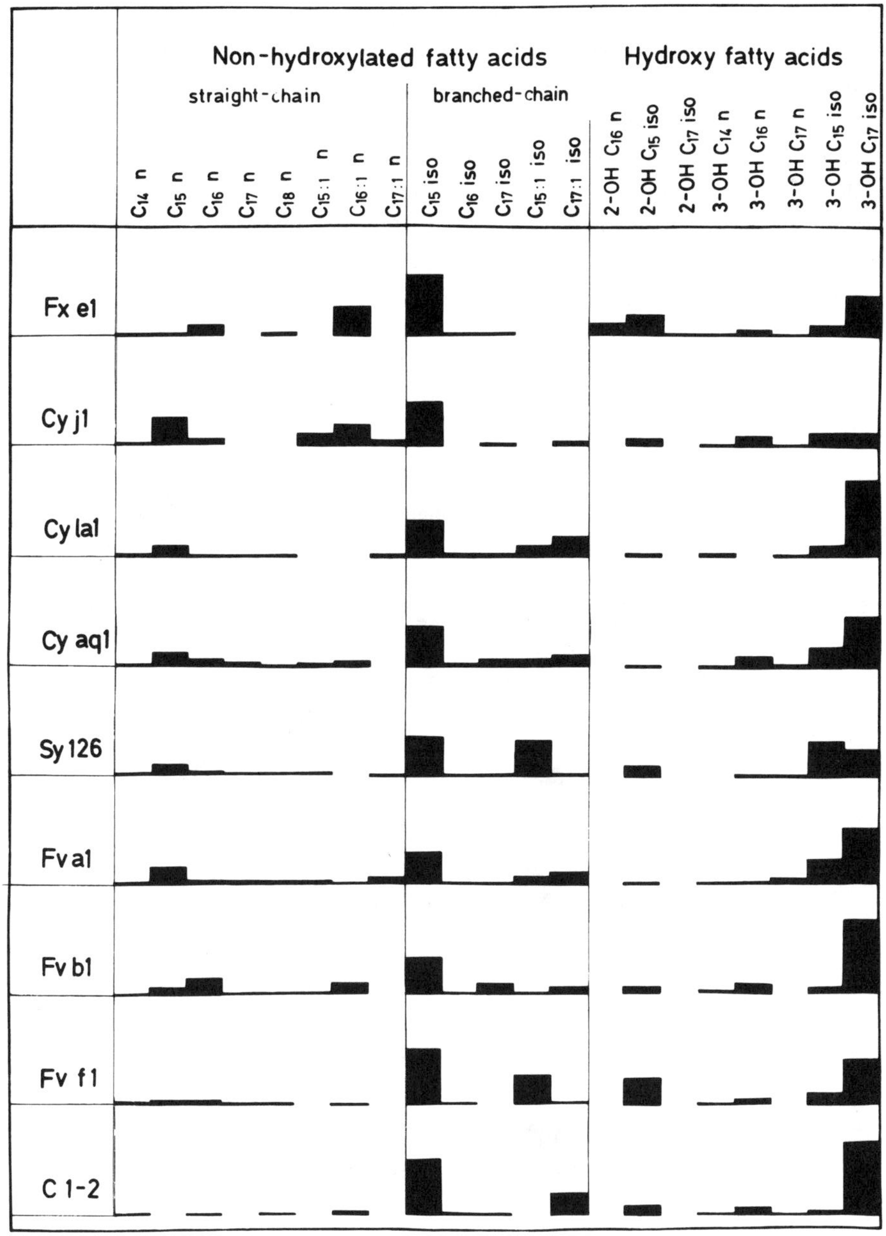


Fig. 1. The fatty acid patterns of Cytophaga-like bacteria (CLB) and selected flavobacteria. Fx e1 = Flexibacter elegans, Cy j1 = Cytophaga johnsonae, Cy la1 = Cy. latercula, Cy aq1 = Cy. aquatilis, Sy 126 = unnamed marine CLB, Fv a1 = Flavobacterium aquatile, Fv b1 = Fv. breve, Fv f1 = Fv. ferrugineum, C 1/2 = unnamed flavobacterium /5/. Bars.: 1 mm = 4.5 %.

The flavobacteria investigated were heterogeneous with respect to their fatty acid patterns. One group of strains had fatty acid patterns very similar to those of the CLB just discussed (Fig. 1). This group includes Flavobacterium aquatile Fv a1, Fv. ferrugineum Fv f1, and several unnamed flavobacteria from Dr. Hayes in Leeds (strains B 4/2, C 1/2, C 20/1, L 10/1 /5/). All these organisms also resemble the CLB in having DNA with a low GC-content (demonstrated for all but B 4/2) /1, 2, 6, 14/, and in containing menaquinones as respiratory quinones /2; as far as Fv a1 is concerned, see addendum to the article of Dr. Mannheim in this volume/. And at least C 1/2 produces flexirubin-type pigments, as are found in many CLB /8/.

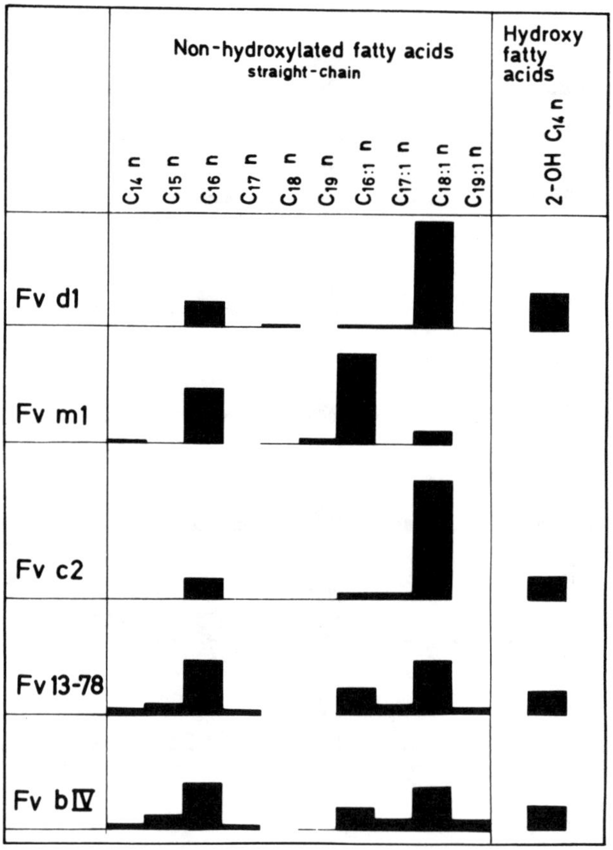


Fig. 2. The fatty acid patterns of selected "flavobacteria". Fv d1 = Flavobacterium devorans, Fv m1 = Fv. meningosepticum, Fv c2 = Fv. capsulatum, Fv 13-78 and Fv bIV = yellow coryneform bacteria /4/. Bars: 1 mm = 4.5 %.

A second group of flavobacterial strains had fatty acid patterns completely different from those described above (Fig. 2). In this group no branched fatty acids were found at all. The patterns of 4 strains resembled one another: the main compounds were $C_{18:1}n$ (up to 60 %) and $C_{16}n$ (10 to 27 %), and the only hydroxylated fatty acid was 2-OH $C_{14}n$ (12 to 20 % of the total). The pattern of Fv. meningosepticum Fv m1 differed markedly: the main components were $C_{16:1}n$ (48 %) and $C_{16}n$ (27 %), and hydroxy fatty acids were totally absent.

As is known from other studies, the latter organisms are taxonomically heterogeneous. Fv. devorans (GC-content 69 %) has just been transferred to the genus Pseudomonas as Ps. paucimobilis /15/. Incidently, the 2-OH $C_{14}n$ acid has been described from this organism by the same authors /15, 16/. Fv. capsulatum (GC-content 63 %) has been shown through the technique of DNA- rRNA hybridization not to be closely related to other flavobacteria, but rather to the pseudomonads (Bauwens and De Ley, in this volume). Strains Fv 13-78 and Fv b IV, obtained from Dr. W. Loeffler in Tübingen and isolated in his laboratory from tap water, are gram-positive and are now regarded as coryneforms /4/. Only Fv. meningosepticum can still be classified among bona fide flavobacteria. It has a GC-content of 37 %, is a menaquinone organism /2/, but its fatty acid pattern is different from all flavobacteria or flavobacteria-like organisms investigated by us.

It thus seems that fatty acid patterns do indeed provide useful and easily accessible information for the classification of gram-negative yellow pigmented bacteria. In addition, our data support the idea that there may be taxonomical connections between certain flavobacteria and the CLB.

The myxobacteria clearly form two groups with respect to their fatty acid patterns (Fig. 3). These groups correspond exactly to the two proposed suborders, the Cystobacterineae and the Sorangineae, which can be distinguished on the basis of cell and myxospore morphology, swarm pattern, and the chemistry of the excreted slime material /7/. The outstanding difference in the respective fatty acid patterns is the total absence of hydroxylated fatty acids in the Sorangineae (Na e1, So ce14). The representatives of the Cystobacterineae contain appreciable amounts of hydroxy fatty acids (15 to 20 %), mainly as the 2-OH $C_{17}$ iso acid. Other differences are less spectacular, e.g. the occurrence of the $C_{17:1}n$ and $C_{18:1}n$ acids in the Sorangineae. All patterns are dominated by branched chain fatty acids, as was to be expected from earlier studies /9, 10, 13/.

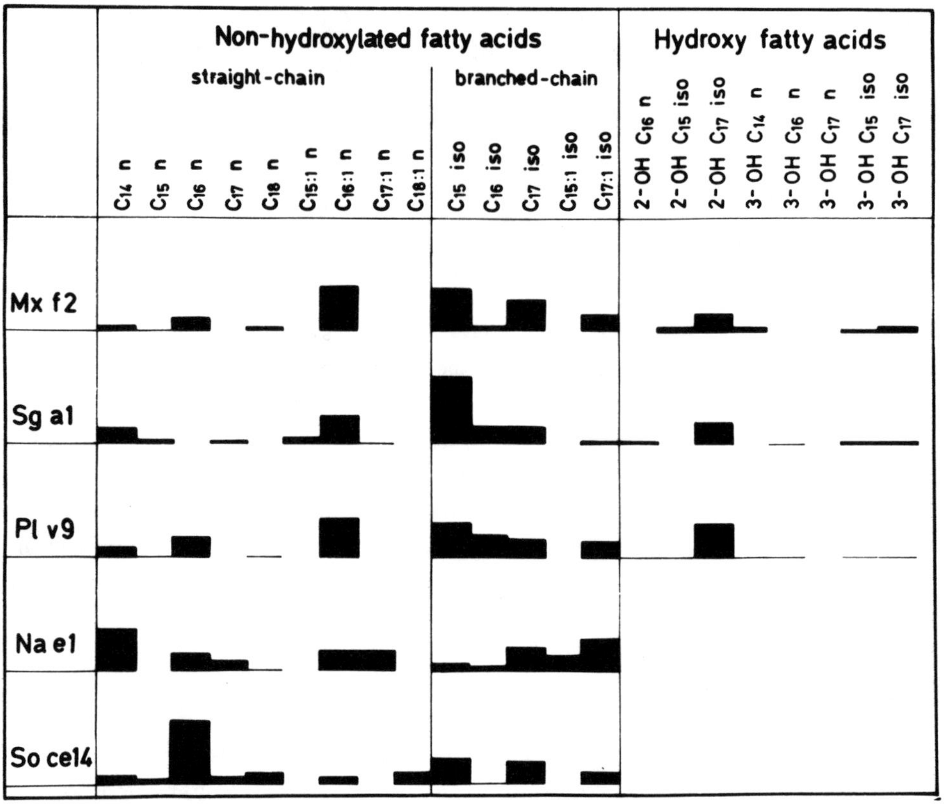


Fig. 3. The fatty acid patterns of myxobacteria. Mx f2 = Myxococcus fulvus, Sg a1 = Stigmatella aurantiaca, Pl v9 = Cystobacter velatus, Na e1 = Nannocystis exedens, So ce14 = Sorangium cellulosum. Bars.: 1 mm = 4.5 %.

## References

/1/ Behrens, H.: Charakterisierung der DNA gleitender Bakterien der Ordnung Cytophagales. Doktor Diss., TU Braunschweig, FRG (1978), 107 pp.

/2/ Callies, E., Mannheim, W.: Internat. J. Syst. Bacteriol. 28, 14-19 (1978)

/3/ Fautz, E., Rosenfelder, G., Grotjahn, L.: J. Bacteriol. 140, 852-858 (1979)

/4/ Haupter, G.: Untersuchungen zur Einteilung gelb pigmentierter Trinkwasserbakterien aus Tübinger Bodenseewasser. Diplomarbeit, Univ. Tübingen, FRG (1980), 65 pp.

/5/ Hayes, P.R.: J. Appl. Bacteriol. 43, 345-367 (1977)

/6/ Hayes, P.R., Wilcock, A.P.D., Parish, J.H.: J. Appl. Bacteriol. 43, 111-115 (1977)

/7/ Reichenbach, H.: Biologie in unserer Zeit 4, 33-45 (1974)

/8/ Reichenbach, H., Kohl, W., Böttger-Vetter, A., Achenbach, H.: Arch. Microbiol. 126, 291-293 (1980)

/9/ Rosenfelder, G., Lüderitz, O., Westphal, O.: Eur. J. Biochem. 44, 411-420 (1974)

/10/ Schröder, J., Reichenbach, H.: Arch. Mikrobiol. 71, 384-390 (1970)

/11/ Takacs, B.J., Holt, S.C., Leadbetter, E.R.: Bacteriol. Proceed. 1968, 124

/12/ Walker, R.W.: Lipids 4, 15-18 (1969)

/13/ Ware, J.C., Dworkin, M.: J. Bacteriol. 115, 253-261 (1973)

/14/ Weeks, O.B.: p. 357-364 in: Buchanan, R.E., Gibbons, N.E. (ed.), Bergey's manual of determinative bacteriology, 8th edition. Baltimore (1974): The Williams and Wilkins Comp.

/15/ Yabuuchi, E., Tanimura, E., Ohyama, A., Yano, I., Yamamoto, A.: J. Gen. Appl. Microbiol. 25, 95-107 (1979)

/16/ Yamamoto, A., Yano, I., Masui, M., Yabuuchi, E.: J. Biochem. 83, 1213-1216 (1978)

SULFONOLIPIDS NOVEL IN PROCARYOTES ARE SIGNIFICANT CELLULAR COMPONENTS OF MANY GLIDING BACTERIA

E. R. Leadbetter and Walter Godchaux III

Biological Sciences Group, U-42
The University of Connecticut
Storrs, CT 06268 USA

Summary

At least two different sulfonolipids, of chemical structure not heretofore recognized to occur in procaryotes, have been isolated, characterized and shown to be present in several genera of non-fruiting gliding bacteria. One of the lipids, capnine, is 2-amino-3-hydroxy-isoheptadecane-1-sulfonic acid. Capnine and N-acylcapnine (the acyl group(s) of which appear to be C-15, C-16, and C-17 3-hydroxy moieties in the one organism studied) constitute 10% to 20% of the lipids of *Capnocytophaga*, *Cytophaga*, *Beggiatoa*, *Flexibacter*, *Vitreoscilla* and *Sporocytophaga*, and up to 3% of the cell dry weight. These quantitatively significant cell components may represent an important chemotaxonomic marker of non-fruiting gliding bacteria, for they appear to be absent from the fruiting myxobacters thus far examined, as they are also from *E. coli* and other bacteria which are either non-motile, or motile as a result of flagellar action. The prospect that they may be associated with, or responsible for, motility in non-fruiting, gliding bacteria is raised.

## Introduction

In the course of examining the envelope fraction of a newly characterized group of gram-negative gliding bacteria (Capnocytophaga [8]) isolated from the human oral cavity, and which are often associated with periodontal diseases [9], we noted an unusual substance. Characterization of the material led to the isolation and identification of capnine (Fig. 1), and during assay of this compound using $^{35}S$-labelling procedures, a related compound(s), N-acylcapnine (Fig. 1), was detected. Collectively termed capnoids, these lipids, which are novel in the procaryotic world, are quantitatively significant membrane components of gliding bacteria in several genera. Selected aspects of the structure(s) and distribution of the capnoids are described here.

Figure 1. Structures of Capnoids

$$(CH_3)_2CH\text{-}(CH_2)_{11}\text{-}\overset{HO}{CH}\text{-}\overset{\overset{+}{NH_3}}{CH}\text{-}CH_2\text{-}SO_3^-$$

Capnine

$$(CH_3)_2CH\text{-}(CH_2)_{11}\text{-}\overset{HO}{CH}\text{-}\overset{\overset{H\ O}{N\text{-}C\text{-}R}}{CH}\text{-}CH_2\text{-}SO_3^-$$

N-acylcapnine

## Materials and Methods

### CAPNINE ISOLATION

Capnine's isolation is relatively easy, because it is insoluble in most solvents when it is in the zwitterionic form, whereas it becomes soluble in moderately polar solvents under alkaline conditions when the amino group is deprotonated. Wet packed cells of *Capnocytophaga gingivalis* strain 30N51 (100 grams) were stirred with 400 ml acetone plus 100 ml methanol until well dispersed at which time 55 ml of 17N $NH_4OH$ were added and stirring continued for an additional 30 minutes. Following filtration and washing of the filter cake with acetone/methanol/$NH_4OH$ (5:5:1), insolubles were discarded and water added to the combined solution, this adjusted to pH 5 with acetic acid and the precipitate which formed collected and washed with ethanol, ethanol containing 0.1 N HCl and water and dried *in vacuo*.

### N-ACYLCAPNINE ISOLATION

Wet, packed cells of an N-acylcapnine-rich organism (e.g., a marine *Cytophaga*) were extracted for lipids using standard chloroform/methanol/water procedures. Insolubles were discarded and water, chloroform, and salt added to isolate the chloroform phase. Hydroxylapatite was used to adsorb the N-acylcapnine from the chloroform solution and after the hydroxylapatite was washed with chloroform and methanol, adsorbed lipids were eluted with chloroform/acetic acid. DEAE-cellulose was then used to adsorb N-acylcapnine, and after washing with acetic acid the lipids were eluted with chloroform/methanol/$NH_4OH$/$NH_4Ac$ and the resultant chloroform solution washed with water to desalt. N-acylcapnine was collected after drying.

### DETERMINATION OF CAPNOID CONTENT

Cells were grown in a medium containing $^{35}S$ sodium sulfate or (for organisms which did not incorporate inorganic sulfate into protein) $^{35}S$ amino acids. Cells were harvested, washed, the pellets collected and weighed. The specific activity of $^{35}S$ sulfate in the growth medium determined by a barium precipitation method. Cells were washed with 5% trichloroacetic acid in water to remove soluble labelled compounds; less than 5% of the labelled capnoids are lost in this procedure. Washed cells were extracted twice with acetone/methanol/ammonium hydroxide (over 99% of the capnoids were extracted by this procedure). Aliquots of the extract were counted for radioactivity and other aliquots chromatographed to determine what proportion of the extracted radioactivity was in capnoids.

From the resulting value for capnoid radioactivity capnoid content was calculated by one of several methods: A) for organisms grown on $^{35}S$ sulfate as the only sulfur source, the content was calculated as radioactivity in capnoids/unit mass divided by $^{35}S$ sulfate specific activity. B) for organisms that can utilize $^{35}S$ sulfate but also require complex growth medium containing organosulfur compounds, a sample of pure capnine was isolated from the lipid extract (after hydrolysis of N-acylcapnine if necessary) and its specific radioactivity determined. Capnoid content was calculated as radioactivity in capnoids/unit mass divided by capnine specific activity. C) growth conditions were as in (B), but isolation of capnine is omitted and the content is calculated as in (A); this provides a minimum estimate as most organisms can utilize both inorganic sulfate and organosulfur compounds for capnoid synthesis. D) as in (C) but contents were calculated on the assumption that all *Capnocytophaga* (grown in the same medium) derive 44% of their capnoid sulfur from inorganic sulfate and 56% from organic sources, as has been observed for strain 30N51. E) *Myxococcus virescens*, which will not incorporate $^{35}S$ inorganic sulfate into protein or capnoids, was grown on medium containing $^{35}S$ amino acids and non-radioactive tryptone. The "effective sulfur specific activity" was calculated from the specific radioactivity of cellular protein (assumed to be 0.7% sulfur by dry weight).

## CHROMATOGRAPHY

Capnoids were chromatographed on Silica Gel G thin layer plates and developed in chloroform/methanol/7N $NH_4OH$ (70/30/5, v/v/v) or in n-propanol/17N $NH_4OH$ (4/1, v/v). Capnine can be detected using 0.5% ninhydrin in butanol (yielding a rose-colored spot) or with n-butanol or water (an opaque zone). N-acylcapnine is detected well with rhodamine, but only poorly with $I_2$ vapor, and is not detected with ninhydrin.

## CELLULAR LOCALIZATION

Cells of *Capnocytophaga gingivalis* were grown in medium containing $^{35}S$ sulfate, harvested, suspended in dilute buffer and divided into two portions one of which was sonicated and the other not. Both were made 70% (w/v) in sucrose, overlaid with sucrose solutions of 55% and 42% and centrifuged to buoyant equilibrium. The gradients were fractionated, protein determined by an assay for dye binding, and radioactivity in trichloroacetic acid-insoluble material determined by a Millipore filter assay.

## Results

### CHEMISTRY

On the basis of its elemental analysis (C, 57%; H, 10.2%; N, 4.2%; O, 18% and S, 9.1%), an apparent molecular weight of 351 deduced from mass spectrometry and the sulfur content, the isolated capnine appeared to be at least 98% pure. The ninhydrin-positive zones seen on thin layer chromatography were coincident with the radioactivity of capnine purified from cells grown in the presence of $^{35}S$ sulfate. Chromatograms sprayed with aqueous rhodamine 6G, exposed to $I_2$ vapors, or charred with $H_2SO_4$ did not reveal the presence of any impurities.

The presence of N as a primary amino group was indicated by quantitative ninhydrin studies in which the color yield of capnine was 85% of that of alanine on a per mol N basis. In as much as inorganic sulfate was not released from capnine after either acid hydrolysis or saponification and capnine's chromatographic behavior was not altered by these treatments we judge that amide and ester groups are absent. The infrared spectrum of capnine (to be described in detail elsewhere [5]) reveals absorption bands readily interpretable as indicating the presence of amino and sulfonate groups; these features are essentially identical to those of the spectrum of taurine (2-aminoethanesulfonic acid). No bands characteristic of double bonds were observed. Capnine was oxidized by sodium periodate to yield a substance that gave a color reaction with the Schiff reagent and had the chromatographic properties of a long-chain fatty aldehyde (the linkage between carbons bearing vicinal hydroxyl and amino groups is oxidized to yield aldehydes). Oxidation of capnine with periodate and permanganate yielded a fatty acid which, when converted to the methyl ester, was positively identified as methyl *iso*pentadecanoate by gas chromatography and mass spectrometry. Together with the elemental analysis and the composition obtained from mass spectrometry, these results establish the presence of the hydroxyl group on carbon 3.

The remainder of the structure of capnine (Fig. 1) was deduced from data obtained on high-resolution mass spectrometry. The composition of the molecular ion was shown to be $C_{17}H_{37}NO_4S$. A fragment with the composition $C_2H_6NO_3S$ (the taurine "head" group) was observed, as was a species from which had been cleaved the group $CH_2SO_3H$; together, these observations establish the locations of the sulfonate and amino groups on carbons 1 and 2, respectively. Evidence was obtained for the linearity of the hydrocarbon chain other than at its terminus. The observation of species from which had been removed one- and three-carbon fragments, but not two-carbon fragments, confirmed the *iso* structure.

When the putative N acyl capnine was examined for infrared absorption, the spectrum showed typical amide bands instead of the primary amine bands characteristic of capnine. After acid hydrolysis the sulfur of N-acylcapnine is in a chemical form that exhibits chromatographic behavior identical to that of capnine. Although N-acylcapnine does not react with ninhydrin, its sulfur-containing hydrolysis product does and the isolated product gives the same color yield per unit weight as does capnine in the quantitative ninhydrin reaction.

We have isolated substantial quantities of N-acylcapnine from only a single organism, a marine Cytophaga species. After acid hydrolysis that compound gave approximately equal weights of a fatty acid fraction and one chromatographically identical to capnine. When the fatty acid fraction was converted to methyl ester derivatives combined gas chromatographic-mass spectrometric analysis of the 3 major components found, indicated that they are 3-hydroxy fatty acid methyl esters. Although it is not yet certain whether the fatty acids are normal or branched, the GC-MS studies indicate that fatty acids of 15, 16 and 17 carbon chain length are present in approximately equal quantities.

## CAPNOID CONTENT

As can be seen in Table 1, when a variety of bacteria were surveyed for the presence of capnoids the latter were detected in many non-fruiting gliding bacteria

Table 1. Capnoid Content of Bacteria

| ORGANISM | METHOD (see Materials and Methods) | CAPNOID CONTENT[a] (μmol/g cell wet wt.) | COMPOSITION (percent) CAPNINE | N-ACYL CAPNINE |
|---|---|---|---|---|
| GLIDERS | | | | |
| Capnocytophaga gingivalis (Str. 30N51) | B | 10.0 ± 0.1 | 85 | 15 |
| Other Capnocytophaga strains | D | 7.2 to 16 | 4 to 85 | 15 to 96 |
| Marine Cytophaga sp. (Str. 7B) | B | 10.0 ± 0.1 | <1 | >99 |
| Beggiatoa alba | A | 8.0 | >99 | <1 |
| Cytophaga johnsonae | C | >3.7 ± 0.3 | <1 | >99 |
| Vitreoscilla stercoraria | C | >3.3 ± 0.1 | <1 | >99 |
| Flexibacter sp. (Str. FS-1) | A | 3.2 ± 0.2 | <1 | >99 |
| Sporocytophaga myxococcoides | A | 2.4 ± 0.2[b] | <1 | >99 |
| Myxococcus virescens | E | ≤0.3 | -- | -- |
| Polyangium cellulosum | A | <0.01 | -- | -- |
| NONGLIDERS | | | | |
| Escherichia coli (Str. K12) | A | <0.01 | -- | -- |
| Mycobacterium-Nocardia sp. | A | <0.01 | -- | -- |
| Sphaerotilus natans | A | <0.01 | -- | -- |

a - Mean ± S.D. of at least two independent determinations, except for Beggiatoa.

b - May represent an underestimate, since extracellular products contribute significant mass to the cell pellet but were apparently absent (or very rare) in fruiting myxobacters such as Myxococcus and Polyangium. Whether they are present in organisms such as Herpetosiphon and Flavobacterium odoratum is not known because these organisms unlike Cytophaga, Vitreoscilla, etc. will not incorporate $^{35}SO_4$ into protein when growing in a complex medium. Such organisms will have to be examined, as was Myxococcus, by examining cells grown in the presence of $^{35}S$ containing amino acids.

CELLULAR LOCALIZATION

An examination of the location of capnoids in cells of gliding organisms has been carried out (only for Capnocytophaga gingivalis) by means of buoyant equilibrium centrifugation in sucrose gradients. Intact cells were shown to band at the 70%/55% sucrose interface, in which fractions all of the cellular protein and capnoids were found. The envelope fractions from sonicated cells banded at the 55%/42% sucrose interface and the fractions here contained most of the capnoids ($^{35}S$) but only a small proportion of the cellular protein; most of the latter remained in the 70% sucrose region (soluble protein). The membranes in the fractions that floated to the 55%/42% sucrose interface were seen by electron microscopic examination to have the form of vesicles surrounded by typical envelope structures (two concentric membranes with intervening "peptidoglycan" material); these vesicles however were devoid of the intensely staining materials (mostly protein) ordinarily found in intact cells of these and other organisms.

The specific radioactivity of the envelope fraction that was greatly enriched in capnoids was equivalent to a capnine content of 160 mg per gram of envelope protein. The capnoids of C. gingivalis used in this experiment are about 85% free capnine but essentially identical results have been obtained with another strain which contains mostly N-acylcapnine. Depending upon the size of acyl moiety of the N-acylcapnine the capnoid content of envelopes from such organisms might range upwards of 300 mg per gram of envelope protein.

## Discussion

The isolation of the capnoids from various gliding bacteria is of interest for several reasons, not the least of which is the rather unusual presence of a sulfonate group; sulfonates occur relatively rarely in biological systems, and in bacteria have not been reported in these molecular forms previously. The other well-recognized lipid sulfonates reported in bacteria are of the sulfonquinovosyl diglyceride—a glycolipid bearing the sulfonate residue on one of the sugar carbons—which occurs in certain pseudomonads [12] in Bacillus acidocaldarius [7] as well as in plants [3].

Structurally and metabolically most naturally occurring sulfonates are related to cysteic acid (alanine-3-sulfonic acid) and the capnoids are probably no exception to this generalization. Capnoids, however, seem unusual in that they are lipids both in the structural sense and by virtue of their location in the cell envelope. We are aware of only one related lipid and that occurs in the gliding eucaryote Nitzschia alba, a diatom which contains 1-deoxyceramide-1-sulfonic acid, a compound similar to N-acylcapnine but in which the 2-amino-3-hydroxy-1-sulfonic acid moiety has an unbranched $C_{18}$ chain that is 4-5 unsaturated. The N.alba sulfonolipid is a minor component of this organism, for it constitutes only 0.13% of the cell dry weight (0.6% of the cellular lipids) [1, 2], in contrast to the capnoids which may constitute 3% of the cell dry weight and over 10% of the lipids, at least in some gliding bacteria.

Clearly, capnine and N-acylcapnine are analogs of sphingosine and ceramides, respectively, but differ from these well-known lipids in two important aspects: the presence of the sulfonate group and the structure of the long-chain amine. These structural analogies undoubtedly extend to biosynthesis as well. The location of the capnoids in cell membranes points to some potential analogies in function as well.

It has been considered unusual for gram-negative bacteria to contain significant quantities of odd-numbered, branched chain fatty acids but such compounds do occur in large amounts both in non-fruiting as well as fruiting gliding bacteria [4, 6, 10, 11]; these fatty acids, as well as both branched and normal hydroxy fatty acids, were known to occur both in total cell hydrolysates of gliding bacteria and in the isolated lipopolysaccharide cores of some of these organisms. Our findings suggest that in addition to the lipopolysaccharides the capnoids may be another, and perhaps more quantitatively significant form, in which some of these fatty acids occur, at least in the non-fruiting gliding bacteria.

But what may be the functional significance of the capnoids? Clearly their absence from the gram-positive and gram-negative bacteria tested (which represented both non-motile as well as organisms motile by flagella) raises the prospect that these compounds might play a role in gliding motility. Given their localization in the cell envelope, of which they are evidently major components, the capnoids might confer unusual properties upon the cell surface. The branched chain fatty acid moieties would be expected to increase membrane fluidity over that characteristic of membranes with normal fatty acids, for example, and the presence of significant quantities of N-acylcapnine may impart high intrinsic negative charge to the membrane surface; in this respect it is interesting to note that most of the surveyed organisms which contain capnoids have N-acylcapnine(s) as the predominant form(s). The significance of the high content of free capnine in some Capnocytophaga strains and in Beggiatoa is unclear but it does suggest that capnine may function not only as a transient intermediate in the biosynthesis of N-acylcapnine(s). Perhaps an interconversion of capnine and its acylated form plays some role in motility; clearly the properties that capnine would confer upon a membrane would be very different from those conferred by N-acylcapnine. The fact that one non-gliding mutant of Capnocytophaga has a distinctly lowered content of capnoids is exciting, and the object of continued study in our laboratory.

If, indeed, capnoids play some role in gliding motility, then the apparent absence of capnoids from gliding myxobacters might be regarded as another suggestion that there is more than one basis for gliding motility in bacteria. If the hydroxy fatty acids of non-fruiting gliders are found exclusively in capnoids, it will be of interest to ascertain whether the fruiting bacteria contain some unusual lipid which does not contain sulfur in which these fatty acids are localized.

The finding that at least three different chain-length hydroxylated fatty acids are released by acid hydrolysis of N-acyl capnine raises the distinct possibility that more than a single molecular species of N-acylcapnine may exist in a given organism. In as much as the capnine studied by infrared spectroscopy and mass spectrometry was isolated from a single Capnocytophaga species, it seems prudent to consider the possibility that there may well be differences in the hydrocarbon moiety of capnines isolated from different bacteria and hence multiple capnines as well; more detailed studies of capnoid chemistry, then, are needed.

Clearly the presence of a new class of biological molecules, apparently restricted to particular groups of organisms, is always of interest for its own sake even if the relation between structure and function is not immediately apparent. The results of on-going experiments to assess the presence of capnoids in other gliding bacteria such as Chloroflexus, Herpetosiphon, Simonsiella, and mycoplasmas, cyanobacters, spirochetes, as well as additional gliding and non-gliding diatoms, selected protozoa, yeasts and other fungi hopefully will provide enlightenment about the significance of the presence of this unusual class of sulfonolipids in cells.

## Acknowledgments

The excellent technical assistance of Ms. K. Fitzgerald and the financial support of the research described here both by NASA (NSG-7450) and the University of Connecticut Research Foundation is gratefully acknowledged.

## References

/1/ Anderson, R., Livermore, B. P., Kates, M. and Volcani, B. E.: Biochim. Biophys. Acta 528, 77-88 (1978).

/2/ Anderson, R., Kates, M. and Volcani, B. E.: Biochim. Biophys. Acta 528, 89-106 (1978).

/3/ Benson, A. A.: p. 387-394 in: Paoletti, R. and Kritchevsky D. (ed.), Advances in lipid research. vol. 1. New York (1963): Academic Press.

/4/ Fautz, E., Rosenfelder, G. and Grotjan, L.: J. Bacteriol. 140, 852-858 (1979).

/5/ Godchaux, W. and Leadbetter, E. R.: J. Bacteriol., in the press (October, 1980).

/6/ Holt, S. C., Forcier, G. and Takacs, B. J.: Infect. Immun. 26, 298-304 (1979).

/7/ Gangworthy, T. A., Mayberry, W. R. and Smith, P. F.: Biochim. Biophys. Acta 431, 550-569 (1976).

/8/ Leadbetter, E. R., Holt, S. C. and Socransky, S. S.: Arch. Microbiol. 122, 9-16 (1979).

/9/ Newman, M. G. and Socransky, S. S.: J. Periodont. Res. 12, 120-128 (1977).

/10/ Schröder, J. and Reichenbach, H.: Arch. Mikrobiol. 71, 384-390 (1970).

/11/ Ware, J. C. and Dworkin, M.: J. Bacteriol. 115, 253-261 (1973).

/12/ Wood, B. J., Nichols, B. W. and James, A. T.: Biochim. Biophys. Acta 106, 261-273 (1965).

# THE *CYTOPHAGA*-LIKE BACTERIA: A SEARCH FOR KEY CHARACTERS

I. Hirsch, H. Reichenbach

Gesellschaft für Biotechnologische Forschung
D-3300 Braunschweig-Stöckheim, FRG

## Summary

Nearly 800 *Cytophaga*-like bacteria (CLB) were isolated from soil and freshwater. Mainly out of these isolates, 185 strains were chosen for a phenotypic analysis, employing 130 different tests. The aim was to find specific traits that would lead to a reasonable clustering and thus could be used as taxonomical key characters. An apparently reasonable classification was obtained with the following characters: gliding motility,cell morphology, cellular shape-change, presence of flexirubin-type pigments and of certain enzymes (catalase, cytochrome oxidase, lecithinase, urease, phenylalanine deaminase, DNase, starch hydrolysis, and gelatin liquefaction). The characters were weighted and used in the order given above. Obviously, this chosen hierarchy has an influence on the resulting arrangement of the strains. Nonetheless, it was possible to define certain groups quite reliably, such as *Flexibacter* and *Lysobacter*. However with other groups, particularly with subgroups, results were often ambiguous. A comparison with data obtained by DNA/DNA hybridization /4/ suggests, that for the different groups and subgroups of CLB different sets of key charatters have to be applied, or that at least the hierarchy of the characters has to be altered.

Anyone familiar with soil and freshwater *Cytophaga*-like bacteria (CLB) will agree, that the taxonomy of these microorganisms still needs intensive study. Much of the uncertainty in this field is due to the fact that in the past most investigators concentrated on physiologically or ecologically specialized groups of organisms, as cellulose, chitin, agar decomposers, fish pathogens etc. This prevented the development of a broad taxonomical concept for the CLB. More recently disagreement with definitions has also contributed to the problem. Although several extensive research reports have been published on this subject within the past decade /7, 8, 10, 11, 15, 17, 18, 21, 22, 23, 24, 25, 26/, we felt that a comprehensive phenomenological study on a large number of diverse and newly isolated strains would be desirable. For neither was it clear to us, whether the existing taxa of CLB were realistic, nor whether their boundaries were well defined, nor even how many taxa there are. Furthermore, our hope was to find additional key characters, that could be used to identify CLB more reliably. As a high degree of uncertainty is connected with any classification based on morphological and physiological data alone, a parallel study was started on a selection of the same strains applying methods of molecular taxonomy /4/.

Gliding motility and yellow pigmentation, which are considered as essential characters of the CLB /20/, served as criteria for recognition. In preliminary experiments, using available strains of Cytophaga johnsonae and Flexibacter elegans, we ascertained which media would allow the expression of these characters. All samples were then plated on several of the differentiating media (details will be published elsewhere). Every colony type appearing on the plates was examined. All unicellular, non-flagellated, gram-negative organism demonstrating gliding, with or without yellow pigmentation, or which did not glide, but were pigmented, were then isolated and cloned.

Then, 185 strains were chosen for a phenotypic analysis. Of these, 120 were randomly selected from the above-mentioned collection, 28 strains had previously been isolated, 7 came from Dr. Güde /16/, the remaining 30 organisms were kindly supplied by other laboratories. Based on classical diagnostic procedures /3, 5, 12/, on the results of previous taxonomic studies on CLB /7, 9, 17, 22/, and on preliminary experiments that were performed with some randomly selected strains, 130 different tests were chosen to be performed on these 185 organisms. Morphological studies on different media were followed by physiological and biochemical characterization.

Two traits proved of paramount importance: cellular shape-change and presence of flexirubin-type pigments. A peculiar cellular shape-change has been described for two flexibacteria, i.e. for FS1 /29/ and for Fx el /28/. In young liquid cultures the cell population consists almost exclusively of very long thread-cells, up to 50 µm in length. These thread-cells are extremely active, gliding, and bending, flexing and twisting in a whip-like motion. As the cultures age, the thread-cells fragment and the cell population changes into short, often almost coccoidal, immotile rods, 2 to 8 µm long. A different cellular shape-change has been described for Cy. johnsonae Cy j1 /1/. Here, in young, actively growing cultures the cells are uniformely short, only 1.5 to 2.5 µm long. As the culture reaches the stationary growth phase, the cells progressively elongate up to 40 µm and more. The long cells are, however, not agile, and have a rough, somewhat broken appearance. These two forms of cellular shape-change were noticed in the present investigation only in peptone containing media. Flexirubin-type pigments can be easily detected by covering the colonies with KOH solution: their yellow or orange color changes then into purple or brown (= KOH-positive) /28/.

The ability to move by gliding was ascertained in chamber cultures, on peptone agar, in all cases, and 92 % of all tested organisms were found positive. Of these gliding strains, again nearly 90 % displayed a dimorphism as described for Cy j1 (called here shape-change 1; Fig. 1). Another 10 % of the gliders, 15 strains, displayed the dimorphism typical for Fx e1 (= shape-change 2; Fig. 1). With the remaining gliders, no definite cellular shape-change could be observed under our experimental conditions (= shape-change 0). Nor did the 8 % of unicellular, non-flagellated, non-gliding organisms show any form of cellular shape-change.

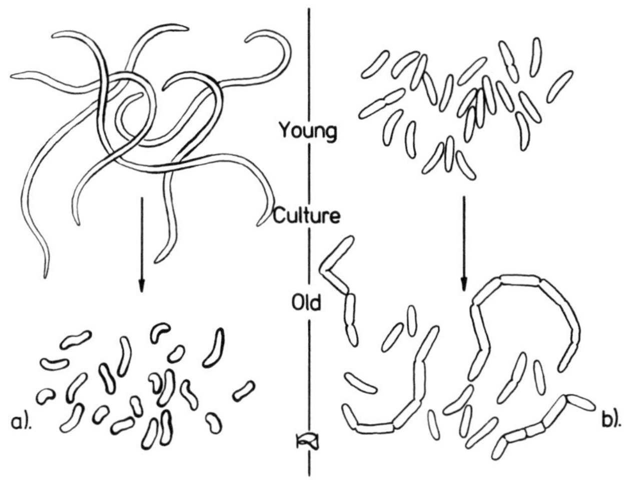


Fig. 1. a) Shape change 2; b) shape change 1.

Approximately 80 % of all tested strains were KOH-positive, which was taken to indicate presence of flexirubin-type pigments. Of these KOH-positive bacteria 95 % simultaneously displayed a cellular shape-change either of type 1 or of type 2. KOH-positive strains were also found among gliders without dimorphism and among non-gliding organisms. KOH-negative strains were noted among gliding as well as among non-gliding bacteria, both types always without a cellular shape-change, however. The scheme below depicts the main groups obtained when the three characters: gliding motility, cellular shape-change, and KOH-reaction are applied to classification.

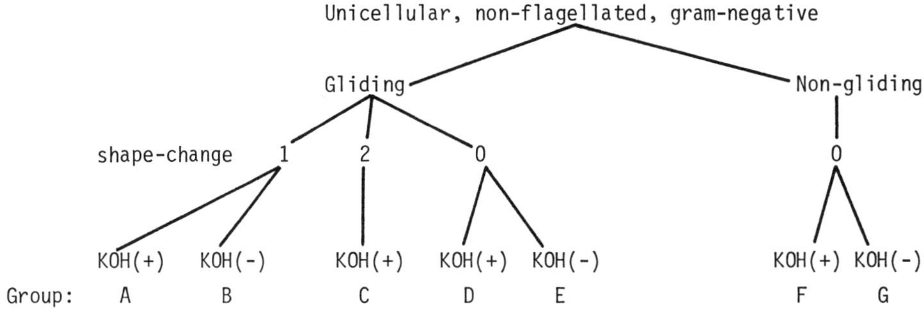

In order to obtain subgroups we then looked for stable morphological and biochemical characters that showed a high degree of correlation. For example, one group of strains formed pale-yellow colonies with smooth, transparent swarm-edges on a certain type of solid medium. On another agar substrate the colonies would be bright yellow and surrounded by a brown zone, due to a water-soluble pigment diffusing into the medium. The same bacteria displayed all the same cellular shape-change and KOH-reaction and could always degrade urea.

The characters thus selected for further classification were: presence of catalase, cytochrome oxidase and lecithinase, breakdown of urea, DNA and starch, production of phenylalanine deaminase, and gelatin liquefaction. These characters were applied in all cases in the same order as given above, and this hierarchy certainly has an influence on the resulting arrangement.

The organisms in group C are characterized by gliding motility, shape-change 2 and a positive KOH-reaction. All of them were furthermore catalase negative, cytochrome oxidase positive, and lecithinase negative. Urea and DNA could or could not be degraded. All 15 strains hydrolyzed starch, and 9 also gelatin. The majority of these bacteria degraded chitin quite readily. This is of particular interest with respect to the proposed new definition of the genus Cytophaga. The GC values were between 44 to 51 % /4/. In our opinion, this group represents the genus Flexibacter /28/.

Group A is characterized by gliding motility, shape-change 1, and a positive KOH-reaction. Here we find 67 % of all organisms tested. Morphologically, these bacteria were slender rods with round or slightly tapered ends. Pigmentation went from pale yellow to deep orange. The majority of the strains was catalase positive, oxidase and lecithinase negative. Each strain degraded at least one kind of complex polysaccharide (chitin, CMC, starch etc.), but none was able to decompose cellulose. The GC values of those strains that have been investigated fell within the range of 32 to 37 % /4/. These organisms we consider as the cytophagas s. str. They can reliably be separated and distinguished from the flexibacters (group B). There is no overlap of the two GC ranges, which gives additional support on the molecular level to the idea that we are dealing with two distinct groups of organisms. We feel, that the differentiating characters mentioned above are better suited to separate the genus Flexibacter from the cytophagas than the single property of polysaccharide degradation /20/. In addition, the group thus defined is essentially in agreement with the original description of at least one species of Flexibacter, i.e. Fx. elegans /cf. 28/.

Returning to group A, several of the subgroups found in the present investigation correspond quite well to clusters obtained by DNA/DNA hybridization /4/. Discrepancies do exist, however, as is illustrated by the following examples.

Example 1: The organisms of one subgroup had been isolated from diverse environments, yet they were quite homogeneous with respect to colony shape, pigmentation and several other key characters. They were the only representatives of group A that were able to degrade urea. Furthermore, all were facultative anaerobes and utilized xylose. DNA/DNA hybridization studies revealed, that one strain, Cy 460, did not belong in this subgroup, for it showed only 40 % homology with the other members, which among themselves were 100 % homologous /4/. Strain Cy 460 had been placed in this subgroup mainly on the basis of urea decomposition and anaerobiosis. Its colony form and pigmentation differed slightly from the remaining representatives of the group.

Example 2: The members of another subgroup were identical with respect to the key characters, but were found to be rather heterogeneous in other traits, such as pigmentation, colony consistency anaerobic growth, $NO_3$-reduction, and chitin degradation. DNA homologies ranged from 45 to 100 %, however, which clearly indicates that this subgroup has to be subdivided /4/.

It is within this cluster that we find strain Cy j1 (NCIB 10150), which is considered a typical Cytophaga johnsonae since it displayes the species specific properties, as stated in the latest edition of Bergey's Manual: amylolytic, chitin-degrader /20/. According to DNA/DNA hybridization data, strain H456 has to be regarded as Cy. johnsonae, too, as it is 70 % homologous with Cy j1 /4/. However, H456 is negative for the species characters of Cy. johnsonae mentioned above. Obviously, the species Cy. johnsonae is ill-defined.

Group B consists of only one strain, Cy. flevensis Cy fl (= A-34), an agar-decomposer isolated from the IJsselmeer, an old North Sea bay /34/. This organism does not contain flexirubin-type pigments, which is typical for marine CLB, and we feel that Cy fl is actually a marine organism.

Groups D and E contain strains that glide, but do not display a cellular shape-change. All strains of group D came from heavily polluted environments. Their rod-cells were generally rather coarse and somewhat pleomorphic. In several of their characters these strains corresponded with the cytophagas of group A. They produced catalase and cytochrome oxidase, and were KOH-positive. The majority were lecithinase positive, which is in contrast to the usual soil and freshwater cytophagas. The GC-values were between 32 and 36 % /4/. However, the genome length of four strains that have been investigated in this respect, was shorter than that of members of group A, and DNA/DNA hybridization data also suggest that there is no close phylogenetic relationship between the bacteria of groups A and D /4/.

The organisms of group E are KOH-negative. Certain key characters allowed them to be separated into two subgroups. The first one comprises the genus Lysobacter /10/. All members were catalase, cytochrome oxidase, and lecithinase positive, and degraded starch and gelatine. Urease and phenylalanine deaminase were not produced. DNase activity was variable. This strikingly homogeneous group was also clearly distinguishable by the very mucoid colonies, and the frequent production of water soluble pigments. Their isolated position was verified by high GC-values between 66 and 70 % /4/. The members of the second subgroup of group E corresponded in several key characters to certain strains of group A (catalase (+), cytochrome oxidase (-), lecithinase (-), ability to degrade starch and gelatine), but differed from the cytophagas in colony form and pigmentation. Their colonies often showed a greenish hue, and generally failed to spread. Unfortunately, only few GC and no hybridization data are presently available for this subcluster. The GC-content of one strain, Cy H3, was 41 % /4/, i.e. slightly higher than with the cytophagas of group A.

The last two groups, F and G, comprise non-gliding organisms. None of the 13 strains showed a cellular shape-change. Approximately 50 % contained flexirubin-type pigments (group F). The majority of these non-gliding organisms resembled the cytophagas in their key characters. However, they generally required growth factors, and their ability to degrade polysaccharides, except for starch, was practically nil. Nine of these strains came from Dr. Hayes' collection of low GC flavobacteria /17/. In fact, overlapping GC-values, common presence of flexirubin-type pigments, similarities in the fatty acid composition /13/, as well as in the quinone spectra /6/ indicate a phylogenetic relationship between certain flavobacteria and the CLB.

When we attempted to place our new isolates into existing taxa of CLB, we were confronted with difficulties. The reasons for this situation are that 1) original type strains, that would allow a direct comparison, are no longer available; 2) the descriptions of the species are often incomplete and vague, resting sometimes solely on differences in temperature optima or colony pigmentation; 3) there is often a lack of adequate comparative studies, so that the boundaries of the species cannot be recognized. The existing "species" must thus be regarded as very problematical. Still another reason for our troubles emerged, however, from our own studies /cf. also 4/: It turned out that the group of the freshwater and soil CLB is far more complex than has previously been assumed. We feel that we should be careful about assigning specific names to any of our isolates.

The data presented in this article are taken from Hirsch, I.: Beiträge zur Taxonomie der Cytophagales auf Grund morphologischer und stoffwechselphysiologischer Eigenschaften. Doktor Diss., TU Braunschweig, FRG (1980), 142 pp.

## References


/1/ Achenbach, H., Kohl, W., Wachter, W., Reichenbach, H.: Arch. Microbiol. 117, 253-257 (1978)

/2/ Anderson, R.L., Ordal, E.J.: J. Bacteriol. 81, 130-138 (1961)

/3/ Bailey, R.W., Scott, G.E.: Diagnostic microbiology. St. Louis (1974): The C.V. Mosby Company

/4/ Behrens, H.: Charakterisierung der DNA gleitender Bakterien der Ordnung Cytophagales. Doktor Diss., TU Braunschweig, FRG (1978), 107 pp.

/5/ Blazevic, D., Ederer, G.M.: Principles of biochemical tests in diagnostic microbiology. New York (1975): John Wiley and Sons

/6/ Callies, E., Mannheim, W.: Int. J. Syst. Bacteriol. 28, 14-19 (1978)

/7/ Christensen, P.J.: Studies on soil and freshwater cytophagas. Ph. D. Diss., Univ. of Alberta, Edmonton, Alberta, Canada (1973), 471 pp.

/8/ Christensen, P.J.: Can. J. Microbiol. 23, 1599-1653 (1977)

/9/ Christensen, P.J., Cook, F.D.: Can. J. Microbiol. 18, 1933-1940 (1972)

/10/ Christensen, P.J., Cook, F.D.: Int. J. Syst. Bacteriol. 28, 367-393 (1978)

/11/ Colwell, R.R.: J. Gen. Microbiol. 58, 207-215 (1969)

/12/ Crowly, N., Bradley, J.M., Danell, J.H.: Practical bacteriology. London (1969): Butterworths and Co.

/13/ Fautz, E.: Physiologische und biochemische Aspekte der Pigmentsynthese bei Flexibacter elegans Fx el (Cytophagales) sowie Untersuchungen zum Fettsäuremuster von gleitenden Bakterien. Doktor Diss., TU Braunschweig, FRG (1978), 147 pp.

/14/ Fuller, W.H., Norman, A.C.: J. Bacteriol. 45, 565-572 (1943)

/15/ Goodfellow, M., Austin, B., Dickinson, C.H.: J. Gen. Microbiol. 97, 219-233 (1976)

/16/ Güde, H.: Appl. Environ. Microbiol. 39, 756-763 (1980)

/17/ Hayes, P.R.: J. Appl. Bacteriol. 43, 345-367 (1977)

/18/ Hayes, P.R., Wilcock, A.P.D., Parish, J.H.: J. Appl. Bacteriol. 43, 111-115 (1977)

/19/ Hendrie, M.S., Mitchell, T.G., Shewan, H.M.: p. 67-78 in: Gibbs, B.M., Shapton, D.A. (ed.), Identification methods for microbiologists, Part B. London (1968): Academic Press

/20/ Leadbetter, E.R.: p.99-107, 111-115, 120 in: Buchanan, R.E., Gibbons, N.E. (ed.), Bergey's manual of determinative bacteriology, 8th edition. Baltimore (1974): The Williams and Wilkins Comp.

/21/ Lewin, R.A.: J. Gen. Microbiol. 58, 189-206 (1969)

/22/ Lewin, R.A., Lounsbery, D.M.: J. Gen. Microbiol. 58, 145-170 (1969)

/23/ McMeekin, T.A., Patterson, J.T., Murray, J.G.: J. Appl. Bacteriol. 34 (4), 699-716 (1971)

/24/ McMeekin, T.A., Shewan, J.M.: J. Appl. Bacteriol. 45, 321-332 (1978)

/25/ McMeekin, T.A., Stewart, D.B., Murray, J.G.: J. Appl. Bacteriol. 35, 129-137 (1972)

/26/ Mitchell, T.G., Hendrie, M.S., Shewan, J.M.: J. Appl. Bacteriol. 32, 40-50 (1969)

/27/ Reichenbach, H.: Biologie in unserer Zeit 4, 33-45 (1974)

/28/ Reichenbach, H., Kleinig, H., Achenbach, H.: Arch. Microbiol. 101, 131-144 (1974)

/29/ Simon, G.D., White, D.: Arch. Mikrobiol. 78, 1-16 (1971)

/30/ Soriano, S.: Ann. Rev. Microbiol. 27, 155-170 (1973)

/31/ Soriano, S., Lewin, R.A.: Antonie van Leeuwenhoek 31, 66-80 (1965)

/32/ Stanier, R.Y.: J. Bacteriol. 53, 297-315 (1947)

/33/ Stanier, R.Y.: p.854-891 in: Breed, R.S., Murray, E.G.D., Smith, N.R. (ed.), Bergey's manual of determinative bacteriology, 7th edition. Baltimore (1957): The Williams and Wilkins Comp.

/34/ Van der Meulen, H.J., Harder, W., Veldkamp, H.: Antonie van Leeuwenhoek 40, 329-346 (1974)

# CLINICAL SIGNIFICANCE AND ANTIMICROBIAL SUSCEPTIBILITY OF FLAVOBACTERIA*

A. von Graevenitz

Department of Medical Microbiology, University of Zürich
CH-8028 Zürich, Switzerland

## Summary

Flavobacteria are not frequently isolated from patient material. F. meningosepticum is a well-known agent of neonatal meningitis which has a poor prognosis. Colonization of compromised hosts in the respiratory tract may occur under epidemic situations. Clinically significant isolates from other sources are in the minority. F. odoratum and II-F have been most frequently isolated from the urogenital tract, and II-J from dog or cat bites. With the exception of the penicillin-sensitive II-F and II-J, most flavobacteria are resistant to many antimicrobials. Effective compounds include rifampin, nalidixic acid, novobiocin, and sulfamethoxazole-trimethoprim. There may be considerable discrepancies between the results of disc and tube dilution tests.

## Introduction

This review supersedes an earlier one published by the author in 1978 which covered the literature until 1977 /96/. Another more extensive review was published by Berger in 1978 /8/. An attempt was made to cover as much of the literature as possible. For a discussion of the sources of flavobacteria, articles appearing elsewhere in this volume should be consulted.

## Frequency of Flavobacteria in Human Specimens

The experience of all authors who have encountered flavobacteria is that

*Prof. Dr. Theo Lammers on his 60th birthday

they are rare isolates from human material occurring in 0.1 % to 1 % of all positive blood cultures /55, 59/; in 0.01 % of all positive urine cultures /55/, in 3 of 4430 /10/ and in 88 of 27600 /67/ genital tract specimens. Epidemic situations constitute exceptions. Environmental samples contain flavobacteria much more often than clinical samples /65/.

## Animal Pathogenicity

F. meningosepticum is not pathogenic for mice or rabbits i.v. or i.p. or for hamsters s.c. or i.v.; intracerebral pathogenicity for mice is doubtful /13, 35, 44/. F. odoratum also shows no animal pathogenicity /86/; and the inference of mouse pathogenicity for F. breve made from the isolation of the organism from canal water /60/ is probably invalid.

## Neonatal Meningitis due to F. meningosepticum

This is the best-known instance of Flavobacterium pathogenicity. Since the first description of the disease in 1944 /82/, around 100 cases have been published /2, 3, 11, 13, 14, 16, 17, 19, 25, 26, 27, 30, 31, 37, 38, 44, 46, 47, 50, 52, 53, 61, 70, 73, 74, 76, 77, 81, 83, 88, 90, 92, 93, 94, 100, 102, 104/. More than half of the infants reported were of low birth weight (< 2500 gm); the same percentage showed an onset of symptoms at the age of less than 8 days (the total range was 1 - 60 days; and reports of 6, 7 /77/ and 19 /104/ months old patients nosologically seem to belong to this group). The most frequent serotype found was C (which was also involved in several epidemics), followed by B, F, A, and E. Symptoms were typical of neonatal meningitis /56/ (failure to feed, respiratory distress, convulsions or lethargy, failure to gain weight, discrete localizing signs). The spinal fluid showed polymorphonuclear pleocytosis, increase in the protein and decrease in the glucose levels, and a positive direct smear. Blood cultures were often positive as well; and serum agglutinins rose in some instances. Few hematological data are extant. The overall case-fatality rate was about 55 %; and in most of the survivors a hydrocephalus developed. Ventriculitis has been blamed for this as well as for the tendency to relapse /92/. Treatment correlated with survival only if intrathecal (and occasionally intravenous /37/) antibiotics were administered whose concentrations in the spinal fluid surpassed the minimal inhibitory concentrations (MICs) of the isolates; e.g. rifampin /47, 76, 92/, sulfamethoxazole-trimethoprim /46/, erythromycin /52, 70, 76/, and vancomycin /31, 37, 70/ although MICs for the latter two drugs were often in the "resistant" range (see "Antimicrobial Susceptibility"). Evaluation of treatment results, however, is clouded by the fact that most infants had received several antibiotics by various routes.

Whether the causative organisms had an environmental or a maternal origin is unknown. In one epidemic /14/ and possibly another one /70/ identical serotypes were isolated from environmental sources, i.e. a faulty sink trap /14/, or saline solution /70/. On the other hand, maternal complications of delivery were reported in only a few instances, and in cases of multiple births only one infant fell ill (8). The likely portal of entry is the pharynx.

Adult Meningitis due to Flavobacterium sp.

Only six cases are extant /5, 44, 46, 53, 54, 76/. They have occurred in patients between the ages of 19 and 69. All of them had underlying diseases predisposing to Gram-negative meningitis /15/ (hematological malignancies, squamous cell carcinoma, malnutrition, chronic renal failure) and showed typical signs and symptoms of such a disease. One case was due to F. meningosepticum type A /46/, two to type C /44, 54/ and one involved Flavobacterium II B /5/; the other two F. meningosepticum strains were not typed. Four patients died. All patients had received multiple antibiotics by various routes.

Pneumonitis and Airway Colonization with F. meningosepticum

In one of the nursery epidemics due to F. meningosepticum type C, five of 14 infants were reported to have had pneumonia with bacteremia /14/; two also had meningitis. Four of the five died but signs and symptoms were not described. Colonization of the upper airways in cohorts of newborns with F. meningosepticum meningitis was common /11, 14, 38, 81/; it was more frequent in prematures /81/ and persisted for about 1 1/2 weeks in those who did not receive specific therapy and for 3 weeks in those who received antibiotics /38/. Colonization of personnel was never observed /14, 38/. Nosocomial F. meningosepticum (type D) pneumonia was observed in a tracheotomized 14 year old patient after ampicillin therapy. Chest roentgenogram and sputum smear were in line with the diagnosis. Therapy with chloramphenicol (to which the organism was sensitive in the disc test) was apparently successful /91/. A case of nosocomial pneumonia in an adult on a respirator was reported in 1979 /61/. In another study, a polymyxin B aerosol was administered to the upper airways of 292 patients in a respiratory-surgical intensive care unit with the aim of preventing Pseudomonas aeruginosa infection /29/. Eighteen patients became colonized with (polymyxin-resistant) flavobacteria, and one developed fatal pneumonia from it. A prospective study from the same unit showed that of 2329 patients, 195 (8.4 %) became colonized with a single strain of Flavobacterium IIB during their hospitalization; the strain was also recovered from numerous water sources in the hospital and from drinking water /24/. Thus - in line with observations on other Gram-negative rods - colonization with flavobacteria is more frequent than infection and usually prefers debilitated individuals.

Etiologically indeterminate strains of flavobacteria have been cultured from sputa of debilitated individuals /65, 68, 78, 95/. No clinical data exist on 43 strains of F. meningosepticum type G and 9 strains of Flavobacterium II-B which were isolated from the respiratory tracts of patients in a medical intensive care unit in Strasbourg /74, 75/.

Isolates of Flavobacterium sp. from the Urinary Tract

A few cases have been reported without clinical data /39, 40, 41, 43, 65, 74/. II-F and F. odoratum strains have been most often isolated from urines /21, 39, 41, 45, 90, 95/ but only a few were deemed clinically significant.

Isolates of Flavobacterium sp. from the Gastrointestinal Tract

Early isolates of F. odoratum were made from the stools of individuals with diarrhea due to known causes /87/. One case of spontaneous peritonitis in a cirrhotic due to II-K-2 ("F. multivorum") has recently been reported /20/; there was improvement with gentamicin and carbenicillin. Our own isolates included one strain each of F. odoratum (pure) and II-F (mixed) from peritoneal fluid /95/.

It should be mentioned here that II-J is an organism that has been isolated as the most frequent aerobe from the oral and nasal fluid of dogs /6, 79/. It is thus not surprising that human isolates were made most frequently from dog or cat bites /90/.

Isolates of Flavobacterium sp. from the Genital Tract

The relative occurrence of flavobacteria in the genital tract was mentioned before (see "Frequency of Flavobacteria in Human Specimens"). II-F is an occasional inhabitant of the human cervix /67, 90/; we have isolated II-F and F. odoratum from endometrial specimens in mixed culture /95/.

Ocular Isolates of Flavobacterium sp.

One patient from whose eye a flavobacterium was isolated had purulent conjunctivitis /34/. Endophthalmitis due to F. meningosepticum type E occurred following keratoplasty. The cornea had been pretreated in a neomycin-gramicidin-polymyxin B solution. The strain was resistant to these drugs and also grew from the culture medium used for storage of the cornea /48/. In this connection, it is of interest to note that in one series, flavobacteria were recovered from 10.8 % of 240 bank eyes /71/ but not from normal eyes. One strain each of II-F /95/ and F. breve /40/ from conjunctiva are reported, however.

Isolates of Flavobacterium sp. from Wounds

Flavobacterium strains from wounds or ulcers have usually occurred in mixed culture and have been etiologically indeterminate /33, 34, 39, 41, 68, 90, 95/ or were associated with contaminated hibitane /17/. Two strains of F. odoratum, isolated in pure culture, were deemed significant: one from an amputation site on a foot repeatedly recovered following gentamicin treatment for a P. aeruginosa infection /18/, and one cultured from a gangrenous foot /41/.

Isolates of Flavobacterium sp. from the Blood

In the experience of this author /95/ and of others /36, 55, 59, 65/, flavobacteria isolated from blood cultures are more often etiologically insignificant than significant. In fact, five isolates reported were associated with the use of contaminated chlorhexidine gluconate /17/. There are, however, reports of significant isolates associated with neonatal meningitis (see above), with isoimmunization with anti-Rh /78/, with tooth extraction (mixed culture!) /69/, with implantation of transvenous pacemakers /51/, and with a variety of immunocompromising disorders /25, 49, 89/. In the former series of 20 cases /25/, there were only three fatalities. This relatively benign course was also evident in three other series. Fever and systemic symptoms of Flavobacterium sp. septicemia in three post-cardiotomy patients reacted favorably to seemingly appropriate antimicrobial treatment /9/. Strains identical to the infecting one were cultured from a water bath used for warming donor blood and from residual water in hoses supplying the heat exchanger of the heart-lung machine /9/. Fever as the only sign of F. meningosepticum (type F) septicemia abated under inappropriate chemotherapy /63, 66/; contaminated injectable drugs seemed to have been the source /64/. Fever persisted, however, under appropriate antimicrobial treatment until arterial catheters, infected with Flavobacterium II-B, were removed /84/. The infections could be traced to contaminated syringes used to obtain arterial blood gas specimens. No clinical data are available on type H to L strains isolated from blood in France /74/.

The first isolate of "Flavobacterium" from a case of subacute bacterial endocarditis /80/ turned out to be Pseudomonas cepacia /96/. Two instances of the disease due to F. meningosepticum were superimposed on rheumatic heart disease. Apparent cure resulted from valvular surgery in one, an addict /101/ while the other patient died /103/. Both had been treated with antimicrobials as well.

Antimicrobial Susceptibility of Flavobacterium meningosepticum and Flavobacterium II-B

From the bacteriological data submitted, it can be assumed that most collectives that were tested represented mixtures of F. meningosepticum and Flavobacterium II-B. The earliest systematic study (1964) on 27 strains /72/ yielded results that contrasted to later ones insofar as the majority of strains showed minimal inhibitory concentrations (MICs) in the "sensitive" range (≤ 10 µg/ml) to ampicillin (AM), chloramphenicol (C), kanamycin (K), and tetracycline (TE).

In line with other investigations, resistance (≥ 20 µg/ml) was observed to cephalothin (CF), colistin (CL), penicillin (P), and streptomycin (S), while most strains were sensitive (≤ 1.25 µg/ml) to erythromycin (E)/72/. A 1969 study /98/ found the majority of 13 strains resistant to CF, AM, K, S, TE, C, polymyxin B (PB); as well as to nitrofurantoin (FD) (≥ 100 µg/ml) but sensitive (≤ 100 µg/ml) to nalidixic acid (NA). It seems as if strains tested later were even more resistant. A 1971 paper /4/ on 11 F. meningosepticum strains found all of them resistant (≥ 25 µg/ml) to the ineffective drugs mentioned above, and also to methicillin (DP), cephaloridine, gentamicin (GM), lincomycin (LI), vancomycin (VA), pristinamycin, and also to carbenicillin (CB; ≥ 100 µg/ml) but again sensitive to NA (≤ 50 µg/ml). All were sensitive or moderately so (≤ 10 µg/ml) to E, novobiocin (NB), and rifampin (RA) but only the latter drug was bactericidal to seven of the strains. Sulfamethoxazole (SX) was inhibitory (≤ 40 µg/ml) for all strains and trimethoprim (T) for none (≥ 20 µg/ml) but the 10 + 1 combination inhibited all strains at 5 µg/ml. Resistance to T was also described in 1973 /12/ and synergism with RA in 1977 /28/. Resistance (≥ 64 µg/ml) to GM, sisomycin, tobramycin (TO) and amikacin (AN) was first described in 1976 /23, 58/. Small numbers of strains have been tested against newer cephalosporins /85, 99/ and were found sensitive (≤ 16 µg/ml) to cefoxitin (CX), variably susceptible to cefotaxim, and resistant to cephamandole and cefuroxime; however, the latter two drugs were tested only on GM-resistant strains. Finally, development of higher MICs to SX, CC, VA and E has been observed during therapy of meningitis with i.v. SX, E, and VA /38/.

Kirby-Bauer disc tests /7/ usually show flavobacteria to be sensitive to E, NB, RA, SXT, NA, and VA; variably susceptible to C, TE, and CX /57/ and resistant to most aminoglycosides (with the possible exception of GM), to the polymyxins, to the older cephalosporins, and to the penicillins /2, 8, 31/.

Comparisons between MIC values and Kirby-Bauer "assignments" led to unexpected results. 6 strains of <u>F. meningosepticum</u> plus 22 strains of "flavobacteria" /1/ and 20 strains of Flavobacterium II-B /97/ were tested using customary MIC breakpoints /62/. All strains proved to be resistant to AM, CB, P, DP, CF, C, TE, CL, VA, GM, TO, AN, K, and S. There was variable susceptibility to clindamycin (CC), E (confirmed in /22/) and sensitivity to RA (≤ 1 µg/ml), NB (≤ 4.0 µg/ml), NA,and SXT (≤ 25 µg/ml). A consistently bactericidal effect was found only with RA. In the Kirby-Bauer tests, however, there were strains "sensitive" or "intermediate" to VA, C, TE /97/, GM /1, 97/ and CX /1/. False sensitivities were thus interpreted from Kirby-Bauer data in all strains sensitive by disc to GM, C, and TE since they actually gave "resistant" MICs. In the cases of CC /1, 97/ and E /97/, some disc-sensitive strains showed "intermediate" MICs. Contrariwise, strains sensitive to CX by MIC exceeded those sensitive by disc but that discrepancy disappeared if sensitive and intermediate strains were combined /1/.

Thus, it seems important to perform dilution susceptibility tests in cases of drugs that display sensitivity in the disc test.

<u>Antimicrobial Susceptibilities of F. breve, F. odoratum, II-B, II-J, and IIK-2</u>

So far, comparative studies between disc and dilution tests have not been done in these species. II-F and II-J resemble each other in their disc sensitivity to P, AM, CB, CF, C, TE, NA, E, SXT, CC, their variable susceptibility to S, GM, NB, FD, and their resistance to TO and K. II-J is reported variably susceptible to polymyxin B while II-F is sensitive /6, 32, 67, 79/. IIK-2, F. breve and F. odoratum behave like F. meningosepticum and Flavobacterium II-B except the strains tested seem to be sensitive to E and - in the case of F. breve - to C and TE by tube dilution /18, 21, 39, 40, 41/.

## References

/1/ Aber, R.C., Wennersten, C., Moellering, R.C.: Antim. Ag. Chemother. 14: 483-487 (1978)

/2/ Abrutyn, E., Plotkin, S.: p. 113-123 in: von Graevenitz, A., and Sall, T. (eds), Pathogenic Microorganisms from Atypical Clinical Sources. New York (1975): Marcel Dekker, Inc.

/3/ Agarwal, K.C., Ray, M.: Indian J. Med. Res. 59, 1006-1009 (1971)

/4/ Altmann, G., Bogokovsky, B.: J. Med. Microbiol. 4, 296-299 (1971)

/5/ Bagley, D.H. et al.: Arch. Intern. Med. 136, 229-231 (1976)

/6/ Bailie, W.E., Stowe, E.C., Schmitt, A.M.: J. Clin. Microbiol. 7, 223-231 (1978)

/7/ Bauer, A.W., Kirby, W.M.M., Sherris, J.C., Turck, M.: Am J. Clin. Pathol. 45, 493-496 (1966)

/8/ Berger, U.: Ergebn. Inn. Med. Kinderk. 41, 1-50 (1978)

/9/ Berry, W.B. et al.: J. Thorac. Cardiovasc. Surg. 45, 476-481 (1963)

/10/ Bovre, K., Hagen, N., Berdal, B.P., Jantzen, E.: Acta Path. Microbiol. Scand. 85B, 27-37 (1977)

/11/ Brody, J.A., Moore, H., King, E.O.: Am. J. Dis. Child. 96, 1-5 (1958)

/12/ Bushby, S.R.M.: J. Infect. Dis. 128S, 10-30 (1973)

/13/ Buttiaux, R., Vandepitte, J.: Ann. Inst. Pasteur 98, 398-404 (1960)

/14/ Cabrera, H.A., Davis, G.H.: Am. J.Dis. Child. 101, 289-295 (1961)

/15/ Chernik, N.L., Armstrong, D., Posner, J.B.: Medicine 52, 563-581 (1973)

/16/ Cintado Bueno, C. et al.: Am. Esp. Pediatr. 11, 147-150 (1978)

/17/ Coyle-Gilchrist, M.M., Crewe, P., Roberts, G.: J. Clin. Pathol. 29, 824-826 (1976)

/18/ Davis, J.M. et al.: Med. J. Austr. 2, 703-704 (1979)

/19/ Daoulas-Le Bourdelles, F., Berche, P., Avril, J.L. Véron, M.: Méd. Mal. Infect. 7, 456-459 (1977)

/20/ Dhawan, V.K. et al.: J. Clin. Microbiol. 11, 492-495 (1980)

/21/ Dick, J.D., Wee, S.B. Noe, D.A., Charache, P.: Abstr. Annu. Meet. ASM C206 (1979)

/22/ Di Persio, J., Krafczyk, T.L.: Antim. Ag. Chemother. 14, 274-276 (1978)

/23/ Drasar, F.A., Farrell, W., Maskell, J., Williams, J.D.: Brit. Med. J. 2, 1284-1287 (1976)

/24/ Du Moulin, G.: J. Clin. Microbiol. 10, 155-160 (1979)

/25/ DuPont, H.L., Spink, W.W.: Medicine 48, 307-332 (1969)

/26/ Eeckels, R., Vandepitte, J., Seynhave, V.: Belg. Tijdschr. Geneeskd. 21, 244-256 (1965)

/27/ Eykens, A. et al.: Helv. Paed. Acta 28, 421-425 (1973)

/28/ Farrell, W., Wilks, M., Drasar, F.A.: J. Antim. Chemother. 3, 459-462 (1977)

/29/ Feeley, T.W. et al.: New Engl. J. Med. 293, 471-475 (1975)

/30/ Furuta, Y. et al.: J. Jap. Assn. Infect. Dis. 48, 313-316 (1974)

/31/ George, R.M., Cochran, C.P., Whealer, W.E.: Am. J. Dis. Child. 101, 296-304 (1961)

/32/ Gilardi, G.: Personal communication (1980)

/33/ Gilbert, D.N. et al.: Arch. Environm. Health 26, 125-130 (1973)

/34/ Greaves, P.W.: J. Med. Lab. Tech. 23, 115-118 (1966)

/35/ Grehn, M.: Zentralbl. Bakteriol. Abt. 1 Orig. A 235, 84-89 (1976)

/36/ Gutmann, L. et al.: Méd. Mal. Infect. 7, 482-485 (1977)

/37/ Hawley, H.B., Gump, D.: Am. J. Dis. Child. 126, 261-264 (1973)

/38/ Hazuka, B.T., Dajani, A.S., Talbot, K., Keen, B.M.: J. Clin. Microbiol. 6, 450-455 (1977)

/39/ Holmes, B., Snell, J.J.S., Lapage, S.P.: Intern. J. Syst. Bacteriol. 27, 330-336 (1977)

/40/ Holmes, B., Snell, J.J.S., Lapage, S.P.: Intern. J. Syst. Bacteriol. 28, 201-208 (1978)

/41/ Holmes, B., Snell, J.J.S., Lapage, S.P.: J. Clin. Pathol. 32, 73-77 (1979)

/42/ Holmes, B., Owen, R.J., Weaver, R.E.: Intern. J. Syst. Bacteriol. (in press)

/43/ Karabatsos, N., Herrold, R.D.: J. Urol. 84, 187-190 (1960)

/44/ King, E.O.: Am. J. Clin. Pathol. 31, 241-247 (1959)

/45/ Kohut, P.: Bratisl. Lek. Listy 71, 65-70 (1979)

/46/ Lapage, S.P., Owen, R.J.: J. Clin. Pathol. 26, 747-749 (1973)

/47/ Lee, E.L., Robinson, M.J., Thong, M.L., Puthucheary, S.D.: Arch. Dis. Child. 51, 209-213 (1976)

/48/ LeFrancois, M., Baum, J.L.: Arch. Ophthalmol. 94, 1907-1909 (1976)

/49/ Lewis, J., Fekety, F.R.: Johns Hopkins Med. J. 124, 106-111 (1969)

/50/ Lietz, R., Handrick, W., Steinert, V., Uhlig, H.: Pädiatrie Grenzgeb. 16, 363-370 (1977)

/51/ Ma, P., Delaney, W.E., Grace, W.J.: Critical Care Med. 2, 135-138 (1974)

/52/ Maderazo, E., Bassaris, H.P., Quintiliani, R.: J. Pediatr. 85, 675-676 (1974)

/53/ Madruga, M., Zanon, U., Pereira, G.M.N., Galvao, A.C.: J. Infect. Dis. 121, 328-330 (1970)

/54/ Mani, R.M. et al.: J. Clin. Pathol. 31, 220-222 (1978)

/55/ Matsen, J.M.: Health Lab. Sci. 12, 305-310 (1975)

/56/ McCracken, G.H.: Hospital Practice 1, 89-97 (1976)

/57/ Moellering, R.C., Dray, M., Kunz, L.J.: Antim. Ag. Chemother. 6, 320-323 (1974)

/58/ Moellering, R.C., Wennersten, C., Kunz, L.J., Poitras, J.W.: Amer. J. Med. 62, 873-881 (1977)

/59/ Monteil,H., Heinrich, V., Richard, C.: Med. Mal. Infect.6, 180-184 (1976)

/60/ Mori, R.: Zeitschr. Hyg. 4, 47-54 (1888)

/61/ Nadarajah, M., Tan, T.H.: Singapore Med. J. 20, 293-296 (1979)

/62/ National Committee for Clinical Laboratory Standards: Performance Standards for Antimicrobic Disc Susceptibility Tests (1979)

/63/ Olsen, H.: Dan. Med. Bull. 14, 1-5 (1967)

/64/ Olsen, H.: Dan. Med. Bull. 14, 6-9 (1967)

/65/ Olsen, H.: Acta Pathol. Microbiol. Scand. 75, 313-322 (1969)

/66/ Olsen, H., Frederiksen, W.C., Siboni, K.E.: Lancet 2, 1294-1296 (1965)

/67/ Olsen, H., Ravn, T.: Acta Pathol. Microbiol. Scand. Sec. B. 79, 102-106 (1971)

/68/ Pedersen, M.M., Marso, E, Pckett, M.J.: Am. J. Clin. Pathol. 54, 178-192 (1970)

/69/ Pinter, M., Ivanyi, J.: Br. Med. J. 2, 1555 (1965)

/7C/ Plotkin, S., McKittrick, J.C.: JAMA 198, 662-665 (1966)

/71/ Polack, F.M., Locatcher-Khorazo, D., Gutierrez, E.: Arch. Ophthalmol. 78, 219-225 (1967)

/72/ Reinarz, J., Mays, B., Sanford, J.: Antimicrob. Agents Chemother. 451-456 (1964)

/73/ Remion, B.: Ped. Pol. 48, 887 (1973)

/74/ Richard, C., Monteil, H., Laurent, B.: Ann. Inst. Pasteur 130B, 141-144 (1979)

/75/ Richard, C., Monteil, H., Le Faou, A., Laurent, B.: Méd. Mal. Infect. 9, 124-128 (1979)

/76/ Rios, I., Klimek, J.J., Maderazo, E., Quintiliani, R.: Antim. Ag. Chemother. 14, 444-447 (1978)

/77/ Samb, A., Chiron, J.-P., Denis, F., Sow, A., Diop Mar, I.: Bull. Soc. Méd. Afr. Noire Lang. Frse. 22, 84-91 (1977)

/78/ Sanchis-Bayarri, V., Borras, R., Mari, M.S.: Rev. Clin, Espan. 133, 455-456 (1974)

/79/ Saphir, D.A., Carter, G.R.: J. Clin. Microbiol. 3, 344-349 (1976)

/80/ Schiff, J., Suter, L.S., Gourley, R.D., Sutliff, W.D.: Ann. Intern. Med. 55, 499-506 (1961)

/81/ Seligmann, R., Komarov, M., Reitler, R.: Br. Med. J. 2, 1528-1529 (1963)

/82/ Shulman, B.H., Johnson, M.S.: J. Lab. Clin. Med. 29, 500-507 (1944)

/83/ Solé-Vernin, C., Ulson, C.M., Zuccolotto, M.: Rev. Inst. Med. Trop. Sao Paulo 2, 54-58 (1960)

/84/ Stamm, W.E., Colella, J.J., Anderson, R.L., Dixon, R.E.: N. Engl. J. Med., 292, 1099-1102 (1975)

/85/ Stephens, M., Potten, M., Bint, A.J.: Infection 7, 109-112 (1979)

/86/ Stutzer, M.J.: Zentralbl. Bakteriol. I Orig. 91, 87-90 (1923)

/87/ Stutzer, M.J., Kwaschina, A.: Zentralbl. Bakteriol. I Orig. 113, 219-225 (1929)

/88/ Sughatadasa, A.A., Arseculeratne, S.N.: Br. Med. J. 1, 37-38 (1963)

/89/ Sullivan, N.M. et al.: J. Infect. Dis., 127, 49-55 (1973)

/90/ Tatum, H.W., Ewing, W.H., Weaver, R.E.: p. 270-294 in: Manual of Clinical Microbiology Washington, D.C. (1974) American Society for Microbiology

/91/ Teres, D.: JAMA 228, 732-733 (1974)

/92/ Torronteras-Santiago, R. et al.: Pédiatrie 33, 393-398 (1978)

/93/ Vandepitte, J., Beeckmans, G., Buttiaux, R.: Ann. Soc. Belge Med. Trop. 38, 563-570 (1958)

/94/ Vandepitte, J., Eeckels, R., Seynhæve, V.: Ann. Soc. Belg. Méd. Trop. 45, 57-72 (1965)

/95/ von Graevenitz, A.: Unpublished results

/96/ von Graevenitz, A.: p. 119-153 in: Gilardi, G.L. (ed.), Glucose Nonfermenting Gram-negative Bacteria in Clinical Microbiology. Boca Raton (1978): CRC Press, Inc.

/97/ von Graevenitz, A., Grehn, M.: FEMS Lett. 2, 289-292 (1977)

/98/ Washington, J.A.: Mayo Clin. Proc. 44, 811-824 (1969)

/99/ Washington, J.A.: Mayo Clin. Proc. 51, 237-250 (1976)

/100/ Watson, K.C., Krogh, J.G., Jones, D.T.: J. Clin. Pathol. 19, 79-80 (1966)

/101/ Werthamer, S., Weiner, M.: Am. J. Clin. Pathol. 57, 410-412 (1972)

/102/ Yabuuchi, E. et al.: Jpn. J. Microbiol. 14, 241-242 (1970)

/103/ Yamakado, M. et al.: J. Jpn. Soc. Intern. Med. 64, 816-820 (1975)

/104/ Zappulla, F. et al.: Min. Ped. 31, 679-682 (1979)

# OCCURENCE OF FLAVOBACTERIUM MENINGOSEPTICUM. SEROTYPES IN FRANCE

C. Richard, B. Laurent

Institut Pasteur, F-75724 Paris Cedex 15

## Summary

Eighty-five clinical strains of Flavobacterium meningosepticum, mostly isolated in France, were assigned to 12 serotypes (number of strains) : A (1), B (1), C (5), F (5), G (55), H (1), I (1), J (6), K (1), L (1), M (1), N (2) ; rough strains (5).

## Introduction

Between 1973 and 1980, 190 strains of the genus Flavobacterium were identified according to the identification schemes of Holmes et al /2,3/ and Gilardi /1/. The distribution of strains into six Flavobacterium species or groups is shown in Table 1.

Rabbits were immunized against reference strains of F. meningosepticum serotypes (A to F) as described by Owen and Lapage /4/. Hyperimmune rabbit sera were used for F. meningosepticum serotyping.

## Results

In an intensive care unit located in Strasbourg University Hospital, 43 F meningosepticum strains (30 patients) were recovered from tracheal exsudate or sputum from January to May 1978. None of these 43 strains was agglutinated by A to F antisera. A new antiserum against the most representative strain of this outbreak was prepared. The new serotype was named G /5/. The 43 strains from the outbreak and 12 other strains isolated in Strasbourg some years before and after the outbreak belonged to serotype G (Table 2).

For serotyping other strains, which were not slide-agglutinated by A to G antisera, 7 more antisera (H to N) were prepared /6/. The titers of A to N antisera and their antigenic cross-reactions are given in Table 3.

The distribution of serotypes A to N (85 strains), their clinical or environmental source and their geographic origin is shown in Table 2.

## Discussion and Conclusion

The number of strains from Strasbourg is high, because permanent cooperation exists between Pasteur Institute (Paris) and Strasbourg University Hospital Laboratories, and, of course, because of the outbreak mentioned above.

We proposed the following epidemiological markers for *F. meningosepticum* :

- serotype (A to N)
- pigment of colonies
- antibiotype, including resistance or sensitivity to O/129 vibriostatic agent
- carbohydrate oxidation pattern.

For example the epidemiological characters of strains isolated during the Strasbourg outbreak were serotype G ; pale pink pigmented ; resistant to O/129 ; resistant to ampicillin, carbenicillin, cephaloridine, streptomycin, kanamycin, amikacin, gentamicin, tobramycin, chloramphenicol, tetracycline, polymyxine B, trimethoprim, nitrofuran, lincomycin, pristinamycin, intermediate resistant to erythromycin, spiramycin, sensitive to minocycline, sulfamide, nalidixic acid, rifampicin, clindamycin, novobiocin ; producing acid (oxidation) from glucose and mannitol, not from lactose and maltose in MEVAG O/F medium.

Improvement of epidemiological markers for *F. meningosepticum* is our current research project.

Table 1. *Flavobacterium* strains isolated from 1973 to 1980 (Institut Pasteur, Paris)

| | |
|---|---|
| *Flavobacterium meningosepticum* | 85 strains |
| *Flavobacterium* King's Group IIb | 59 strains |
| *Flavobacterium odoratum* | 14 strains |
| *Flavobacterium breve* | 3 strains |
| *Flavobacterium* King's Group IIf | 2 strains |
| *Flavobacterium* sp | 26 strains |

Table 2. Flavobacterium meningosepticum. Serotype distribution

| Serotype | N° | Source | Geographic origin |
|---|---|---|---|
| A | 1 | Blood | Strasbourg |
| B | 1 | Blood | Unknown |
| C | 5 | C S F (4)<br>Catheter (1) | Noumea (2) Paris (1) unknown (1)<br>Unknown |
| F | 4 | Blood (2)<br>Sputum (1)<br>C S F (1) | Brest (1) Strasbourg(1)<br>Limoges<br>Dakar (Sénégal) |
| F,H | 1 | Blood | Leuven (Belgium) |
| G | 55 | Tracheal exsudate (36)<br>Sputum (12)<br>Blood (4)<br>Wound (1)<br>Catheter (1)<br>Respirator (1) | Strasbourg |
| H | 1 | Blood | Strasbourg |
| I | 1 | Blood | Strasbourg |
| J | 6 | Skin (3)<br>C S F (2)<br>Blood (1) | Strasbourg<br>Strasbourg<br>Saint-Denis |
| K | 1 | Blood | Saint-Nazaire |
| L | 1 | Urine | Strasbourg |
| M | 1 | Blood | Copenhagen (Denmark) |
| N | 2 | Sperm (1)<br>Wound (1) | Paris<br>Calais |
| Rough strains | | Wound (3)<br>Blood (1)<br>Catheter | Bordeaux (2) Lille (1)<br>Copenhagen<br>Copenhagen |

Table 3. Agglutination tests

| Strains | Antiserum to F. meningosepticum serotypes | | | | | | | | | | | | | |
|---|---|---|---|---|---|---|---|---|---|---|---|---|---|---|
| | A | B | C | D | E | F | G | H | I | J | K | L | M | N |
| Serotype A | 640 | - | 20 | - | - | - | 20 | - | - | - | - | - | - | - |
| Serotype B | - | 320 | - | - | - | - | - | - | - | - | - | - | - | - |
| Serotype C | - | - | 640 | - | - | - | - | - | - | - | - | - | - | - |
| Serotype D | - | - | - | 160 | - | - | - | - | - | - | - | - | - | - |
| Serotype E | - | - | - | - | 160 | - | 20 | - | - | - | - | 10 | - | - |
| Serotype F | - | - | - | - | - | 320 | - | - | - | - | - | - | - | - |
| Serotype G | - | - | - | - | 20 | - | 320 | - | - | - | - | - | - | 10 |
| Serotype H | - | - | - | - | - | - | - | 320 | - | - | - | - | - | - |
| Serotype I | - | - | - | - | - | - | - | - | 160 | - | - | - | - | - |
| Serotype J | - | - | - | - | - | - | - | - | - | 160 | - | - | - | - |
| Serotype K | - | - | 40 | - | - | - | - | - | - | - | 80 | - | - | - |
| Serotype L | - | - | - | - | - | - | - | - | - | - | - | 80 | - | - |
| Serotype M | - | - | - | - | - | - | - | - | - | - | - | - | 20 | - |
| Serotype N | - | - | - | - | - | - | - | - | - | - | - | - | - | 1280 |

## References

/1/ Gilardi, G.I. (ed), Glucose non fermenting gram negative bacteria in clinical microbiology, p 51-57, West Palm Beach, Florida (1978) : CRC Press.

/2/ Holmes, B., Lapage, S.P., Malnick, H. : J. Clin. Pathol. 28, 149-155 (1975).

/3/ Holmes, B., Snell, J.J.S., Lapage, S.P. : Int. J. Syst Bacteriol. 27, 330-336 (1977).

/4/ Owen, R.J., Lapage, S.P. : Antonie van Leeuwenhoek. 40, 255-264 (1974).

/5/ Richard, C., Monteil, H., Le Faou, A., Laurent, B. : Méd. Mal. Infect. 9, 124-128 (1979).

/6/ Richard, C., Monteil, H., Laurent, B. : Ann. Microbiol. (Inst. Pasteur). 130B, 141-144 (1979).

# THE SLOW GROWING GRAM NEGATIVE PIGMENTED WATER BACTERIA

L.G. Herman, BSA, Ph.D.

National Institutes of Health
Environmental Safety Branch
Bethesda, Maryland 20205

Summary

Bacteria, producing yellow to orange pigmented colonies on the surface of natural and artificial media, can be found in virtually every segment of the environment. They can be isolated from the soil, from fresh and salt water, from rain and snow, as well as from man, animals, birds, fish and plants. The majority of pigmented organisms are slow growing and can best be seen on the surface of agar plates and membrane filter pads, but usually only after extended incubation of 5-15 days at room temperatures. Some species appear to be chlorine tolerant and can be readily isolated from domestic water supply outlets such as taps, spigots and drinking fountain heads. They also grow readily in static water holding units such as tanks, water baths and special equipment, commonly noted in laboratories and hospitals. Although most of the yellow pigmented bacterial colonies can be classified as harmless microorganisms, occurring in most natural water supplies; infections can and do occur among infants, debilitated patients, surgical patients and patients on immunosupressive drug therapy, chiefly through careless handling of water supplies contaminated with these organisms.

The use and abuse of our natural fresh water supplies is an ever increasing worldwide problem. Not only are the lakes and streams polluted by chemicals, fertilizers, and insecticides from surface waters, but the sewage disposal practices encourage the survival and growth of many undesirable species of microbes. Bayliss /2/ in 1930 foresaw the coming events when he stated, "It is easy to remove microorganisms and avoid other particles by filtration, but it is not easy to reduce the organic content of many waters to the point where it will not support microbial growth." The truth of this statement was verified many times over a period of twenty years of microbiological analyses of water samples in a hospital and research facility.

The samples came from many sources such as dynamic and free flowing and static or stored water supplies especially in dead end or little used pipe lines, water baths, reservoirs, tank and storage units. Most samples at one time or another, contained viable mold spores, yeast cells, acid fast bacilli, aerobic and anaerobic spore formers, micrococci, streptococci and diptheroids in addition to many species of pigmented and nonpigmented gram negative rods /14/—some of which still remain to be classified as this symposium is attempting to demonstrate. Our attention was first drawn to this problem when four open heart surgery patients developed a bacteremia during their convalescent period /4/. Environmental studies demonstrated a similar organism in a water bath for tempering blood in the surgical area, and in the water cooling system of the heart lung machine used on the four patients, as was isolated from blood cultures of these patients. Subsequent environmental studies showed that various species of pigmented gram negative bacteria could be readily isolated from the water from virtually every area, in varying numbers, of the research complex.

Since water is the "almost universal" solvent, it is able to extract both beneficial and harmful components from the soil or surfaces over or through which it flows and percolates that give it flavor, odor, color, hardness, acidity, alkalinity, and the nutrients that support and encourage microbial growth before, during and after routine use and handling. Chambers and Clarke /7/ describe it clearly, "the extent to which water can serve as a bacterial growth medium is one of the least recognized facets of microbiology."

The typical water bacteria, i.e., the pigmented gram-negative, non-fermenting, slow growing species require somewhat different conditions for optimum development. Pigment is most readily produced on the surface of agar plates incubated at room temperature for 5-15 days. The slow growth is not limited to artificial media, since samples of water held at room temperature show a slow but steady increase in numbers, while the levels of coliforms or other non-pigmented species usually decrease during similar holding periods.

## Pigment and Biochemical Characteristics

While it is easy to recognize pigmented colonies on the surfaces of agar plates or membrane filter pads, Gilardi /11/ rightly comments that, "using pigment production as a taxonomic criterion is in itself of dubious merit because of the effects of variation in temperature and media on pigment production." Ciegler /8/ noted that over half of the marine bacteria isolated to date, contain unsaponifiable fat soluble yellow to red pigments. Hayes /12/ separated pigmented cultures isolated from fish into Flavobacteria and Cytophaga species. Starr and Stephens /32/ noted that many phytopathogenic bacterial colonies isolated from plants were also yellow in color. Bean and Everton /3/ isolated 172 cultures of orange-yellow pigmented bacteria from chlorinated cooling waters and classified most of them as Flavobacteria. On the other hand, Koyama et. al. /19/ noted non-pigmented Flavobacteria and concluded that, "pigmentation of these bacteria is so variable that the description is not rigid and they must therefore be considered to be a heterogeneous species." Although Hayes /12/ and Shewan /28/ feel that the Flavobacteria should be classified as Cytophaga , Weeks /33/ states that "Flavobacterium is more a historic concept than a taxonomic reality." However, Gilardi /11/ defines the minimal features required for the identification of Flavobacterium as: (1) Gram negative asporogenous bacilli, (2) polar flagellated, if motile, (3) oxidize carbohydrates to acid without gas, (4) usually indole positive, (5) catalase positive, and (6) indophenol oxidase positive and (7) gelatin positive.

## Isolation Methods

Of the many problems associated with the slow growing, pigmented, gram negative water bacteria, that of original isolation is probably the most frustrating. Although various species of Flavobacteria and Cytophaga, Table I, can grow over a wide range, $4^{o}C$ - $44^{o}C$, of temperatures, most species grow best at room temperature, $18^{o}C$ - $32^{o}C$, but require extended incubation of 5-15 days to produce visible colonies and measurable pigmentation /14/, hence the plates should be kept in plastic bags to prevent undue loss of moisture from the agar base. Since most species are pigmented and aerobic, swabbed or surface inoculated plates are preferable to poured plates, in that the pigments can be more readily noted and colony growth is more

Table I

Slow Growing Pigmented Water Bacteria in Flowing and Static Water Supplies

| Source | No. Positive / No. Sampled[a] | % Positive | Maximum Colony Count/ml |
|---|---|---|---|
| Drinking Fountains 1965 | 117/150 | 78 | $3 \times 10^{5}$ |
| Drinking Fountains 1975 | 164/217 | 76 | $4 \times 10^{3}$ |
| Drinking Fountains 1978 | 334/440 | 75 | $3 \times 10^{3}$ |
| Sink Faucets | 48/69 | 70 | $3 \times 10^{3}$ |
| Dead End Pipe Lines | 10/10 | 100 | $5 \times 10^{4}$ |
| Water Baths | 5/7 | 71 | $7 \times 10^{5}$ |
| X-ray and Photo Wash Tanks | 10/20 | 50 | $2 \times 10^{3}$ |
| Humidifying Units | 36/55 | 65 | $4 \times 10^{7}$ |
| Dental Chair Spray Units | 7/7 | 100 | $2 \times 10^{4}$ |
| Distilled Water Lines | 5/10 | 50 | $3 \times 10^{4}$ |

[a]Dilutions spread on surface of T.S.A. plates and incubated at room temperature 5-15 days.

easily transferred from the agar surfaces. Enrichment methods in broth are not recommended because the slow growing cells are often "lost" in the heavy and more rapid growth of other species of bacteria. However, it has been noted that, in many instances, when lactose broth tubes from M.P.N. determinations of drinking water are held for 5-15 days at room temperature, pigmented growth is readily visible in many gas free tubes and this, when streaked on agar plates, usually also produces pigmented colonies. Since pigmented bacterial colonies can be isolated from water from virtually every corner of the globe, a variety of media /27/ are advisable to provide the proper conditions for optimum growth, pigment production and identification of the various species. Basic nutrient meat or soybean extract agar can be enhanced by the addition of yeast extract, plant proteins and carbohydrates, fish meal, milk casein, as well as soil extract and sea salts, with and without the addition of sheep blood. To prevent overgrowth by gram positive cocci and spore formers, the addition of antibiotics to the media is helpful, since most Flavobacteria are antibiotic insensitive.

Where the microbial contamination is minimal, i.e. less than 1 viable organism in 10, 50 or 100 ml, as in the case of distilled, tap, deionized or reverse osmosis treated water, the membrane filter technique should be used with 50-500 ml samples so that 10-20 visible colonies can develop per pad on extended incubation of 5-15 days at room temperatures.

Table II

Water Samples Membrane Filtered Through 47 mm x 0.45μ Pads[a]

| Source | Samples Tested | Sample Size | Colonies Per Pad |
|---|---|---|---|
| Tap water through clean faucet | 15 | 250 ml | <10 (5)<br>10-200 (10) |
| Tap water through rubber hose | 20 | 250 ml | <200 (6)<br>TNTC[b] (14) |
| Distilled water from still (direct) | 30 | 500 ml | No Growth |
| Distilled water at end of system; clean faucet | 20 | 250 ml | <200 (5)<br>TNTC (15) |
| Distilled water through rubber hose | 20 | 50 ml | <200 (5)<br>TNTC (15) |
| Deionized water, central unit | 5 | 250 ml | TNTC (5) |

[a]Pads laid on T.S.A. plates, incubation at room temperature 5-15 days.

[b]TNTC -too numerous to count.

Since some species of Flavobacteria are photochromogenic, incubation should be continued in subdued light rather than in direct sunlight or total darkness and where the temperatures are not likely to vary greatly /19/.

## Dynamic Water Supplies

### Drinking Fountains

It is amazing that the "drinking fountain head" still remains the most ignored part of the entire water delivery system to the individual in the average community. Table I illustrates the variations and degree of contamination and levels of pigmented water bacteria in one hospital. In 1965 a total of 78% of the drinking fountains were contaminated, yet thirteen years later there was little (75%) significant difference in the number of contaminated units.

Another factor often overlooked when water from drinking fountains is compared with water from sink faucets, is the absence of chlorine in the fountain water. Many public buildings such as schools, hospitals, offices, etc., have a separate water delivery system for drinking fountains which is cooled or chilled, especially in the summertime, and constantly circulated so that the chlorine content usually falls below the effective level of 0.6 ppm, thereby permitting microbial growth and multiplication in the system and especially at the air-water interface on the fountain head. Close examination of many fountains leaves much to be desired in the way of aesthetics. Many have easily visible accumulations of deposited scales, others are heavily rusted, occasional heads have a deposit of slimy growth, often yellowish to brown in color /13/ /14/. Nevertheless, the drinking fountain head, unless routinely brushed and polished, is often little better than the "common cup" that is seldom cleaned or sanitized.

### Distilled, Deionized and Reverse Osmosis Units

The continuing demand for water of high purity in pharmacy, biochemical, microbiological and tissue culture research has been developed to a point where distilled water is now available without demonstrable pyrogens and free from viable bacteria /16/. However, the usual central distilled water supply tank with extended pipe lines and rubber hoses, will always be contaminated with a wide variety of pigmented and non-pigmented organisms /6, 9, 14, 17/.

Deionizing units of various sizes are also able to provide water with very low levels of anions and cations, but microbial contamination remains a problem. Stamm et.al., /31/ found Flavobacterium in every deionized and softened water unit and outlet in their laboratories and noted growth in the resin columns of up to $10^5$ cells per ml. DeRoos et. al., /9/ noted that when both deionized water and distilled water was available in the laboratory, the distilled water had a lower bacterial and pyrogen content than the deionized supply.

A water supply, low in sodium and calcium salts and virtually free from pyrogens and viable bacteria for use in hemodialysis, seems to be possible with a reverse osmosis water production system /15/. However, it was noted that critical maintenance and strict sanitation is necessary to prevent recurring contamination with both pyrogens and viable organisms /25/.

## Static Water Supplies

### Water Baths and Wash Tanks

The water in this equipment, static, open, seldom changed, and held at varying temperatures from ambient room to 45°C provides an ideal medium for microflora of all species (Table III) but even here, *Flavobacterium* seem to be the predominant organism - up to 35% of the recognizable types.

Table III

Predominant Microbial Species Isolated From 100 Laboratory Water Baths, (0°C-45°C)

| | No. Contaminated |
|---|---|
| *Flavobacterium* sp. | 21[a] |
| *Flavobacterium* with mixed flora | 14 |
| *Klebsiella* sp. | 20 |
| Other Gram negative species | 20 |
| *Pseudomonas aeruginosa* | 4 |
| Gram positive spore formers | 6 |
| Yeast species | 3 |
| Sterile - No growth | 12 |

[a] Swabs on T.S. agar plates held 5-15 days at room temperature.

Contaminated water baths are especial hazards in surgical suites where blood packs are tempered /4/ or in dressing rooms where saline transfusions are warmed, and especially in microbiological laboratories when aerosols created by splashing, etc., tend to release contaminants into media, table tops, equipment, etc. The most effective and least hazardous or corrosive inhibitor noted to date is the addition of 0.1% - 0.6% of glutaraldehyde to the water bath. The water should then remain inhibitory for extended periods of time depending on the addition of organic matter through spills, etc. Regular cleaning and the use of distilled water is also recommended.

## Nebulizers and Humidifiers

A critical study of micro organisms /29/ in heated nebulizers showed almost complete control of growth of all types of gram negative rods, but chiefly Pseudomonas, Moraxella, Acinetobacter, and Flavobacterium species when the water temperature was held at 46°C or above while those units held at 41°C or below, showed 38% - 52% contamination when held for 5 days or more before cleaning. It was also noted that tap water, when used in humidifying units, was the source of infection for 5 of 8 infants exposed to these contaminated mists /21/.

## Dental Chair Units

Studies of the water spray from a dental hand piece /1, 20/ (Table I), showed a high level of contamination by pigmented gram negative organisms in every unit checked, which could be a hazard to both patient and dentist, especially since some counts exceeded 5,000,000 organisms per ml. This contamination problem is further complicated by the fact that dental reservoirs are usually heated to 35-37°C to provide a warm spray. The possibility that the pigmented colonies often noted in sputum specimens /26/ may in some cases originate from dental sprays, should not be overlooked.

## Hemodialysis Systems

Favero et. al./10/ noted that water produced by deionization, distillation, reverse osmosis or softening, supported many species of gram negative bacteria especially when stored or used in the dialysate of artificial (hemodialysis) kidney machines. In many instances, the proliferating organisms were often the cause of pyrogenic and septic reactions in the treated patient. They emphasized the hazards of "naturally occurring" bacteria—in which organisms present in distilled water and diluted in the same "membrane filtered" water grew more rapidly, showed higher disinfectant resistance and longer survival times than did those organisms that were passed through conventional laboratory media and then reintroduced into the same waters. Among the nine gram negative organisms usually isolated from dialysis fluids, Flavobacterium, Pseudomonas and Acinetobacter are the most common and troublesome contaminants.

# Infection Hazards with Pigmented Water Bacteria

## Premature and Newborn Infants

With the advent of antibiotics, more attention is being paid to the gram negative bacteria capable of surviving in the presence of many antibiotics at higher than average levels required to inhibit the gram positive species. The emergence of a slightly pigmented gram negative non-motile, oxidase positive bacillus that produced indole but had little reaction on sugars and was sensitive only to erythromycin and carbenomycin, was the apparent cause of spinal meningitis in 19 newborn infants of which fifteen died /5/. King /18/ studied these organisms further and in 1959 separated them into three serological groups, and because of its light yellow pigment and association with meningitis, tentatively classified it as a new species— Flavobacterium meningosepticum. Following this publication, sporadic cases have been noted in the international literature annually /14, 22, 34/.

### Mature Patients Infected Following Surgery

Olsen /22/ isolated F. meningosepticum from ten cases of post operative infection following thoracic surgery in Denmark, all patients survived. To determine the source of these infections, he was able to isolate F. meningosepticum from brook water, garden soil and from the blood, urine, sputum and spinal fluid of various patients—all similar to the 1959 King isolates.

The hazards of indwelling arterial catheters are well described by Stamm et. al., /30/ in which they noted 14 patients with positive blood cultures due to a Flavobacterium. The organisms were isolated from the catheters, stop cocks and ice supply used to cool the syringes. Their recommendations are simple—sterile techniques, handwashing and fresh catheters every 24 hours.

## Discussion

Since the slow growing pigmented water bacteria are usually found in circulating and static water supplies even in the presence of detectable levels of chlorine /13/, it may mean that many are chlorine tolerant and survive to grow and multiply when the chlorine has been dissipated. Although these organisms may be "naturally occurring" in water supplies, /10, 23/ they grow so slowly and better on the surface of agar plates, that they can easily be missed because most water analyses tubes and poured plates are discarded after 48 hours incubation.

While various species of Flavobacterium have been isolated from patients following use of products contaminated by them, viz /11, 23, 24, 30/ hence, one should not overlook the hospital water supplies as a possible source of those infections, even though they are difficult to identify and trace /1, 4, 10/.

Although the average laboratory may not have the personnel or experience available to identify each and every waterborne organism, as has been shown many times during this symposium, they can easily demonstrate the presence of other than coliform species by: (1) swabbing the surface of various types of agar plate media directly, or by (2) using larger amounts, e.g., 50-500 ml of treated water for membrane filtration, both with extended incubation.

The ability to recognize microbial growth in any kind of potable, distilled or deionized water supplies when used for domestic, medical or research uses, thus becomes a responsibility of the microbiologist associated with the consuming unit.

Therefore, depending on the final use of any given water supply, the presence of slow growing, pigmented water bacteria can be (1) harmless indicators of an unsterile condition, or (2) they can be a hazard to the system, the user and the end product. The degree to which this occurs, depends on the pressures created by our changing society on the total environment and eventually on the microflora within it. Then as some species are stimulated or altered by the effect of acids, antibiotics, antibodies, antiseptics, chemicals, disinfectants, gases, petroleum products, etc., their adaptation becomes more easily recognizable as "survivors" in a complex population, which is certainly true of the yellow pigmented gram negative bacteria of the Flavobacterium - Cytophaga group of microorganisms.

## Bibliography


/1/ Abel, L.C., Miller, R.L., Micik, R.E., and Ryge G.: J. Dent. Res. 50, 1567-1569 (1971)

/2/ Baylis, J.R.: Water Works and Sewerage, 77, 335-338 (1930)

/3/ Bean, P.G., and Everton, J.R.: J. Appl. Bact., 32, 51-59 (1969)

/4/ Berry, W. B., Morrow, A.G., Harrison, D.C., Hochstein, H.D., and Himmelsbach, C.K.: J. Thorac. and Cardiovas. Surg. 45, 476-480 (1963)

/5/ Brody, J.A., Moore, H., King, E.O.: Am. J. Dis. Child. 96, 1-5 (1958)

/6/ Carson, L.A., Favero, M.S., Bond, W.W., and Peterson, N.J.: Appl. Microbiol. 23, 863-869 (1972)

/7/ Chambers, C.W., and Clarke, N.A.: Adv. Appl. Microbiol. 8, 105-143 (1966)

/8/ Ciegler, A.: Adv. Appl. Microbiol. 7, 1-34 (1965)

/9/ DeRoos, R., Vesley, D., DuChene, A.G., and Hart, L.J.: Health Lab Science, 13, 11-19 (1976)

/10/ Favero, M.S., Petersen, N.J., Carson, L.A., Bond, W.W., and Hyndman, S.H.: Health Lab Science, 12, 321-334 (1975)

/11/ Gilardi, G.L., Hirschl, S., and Mandel, M.: J. Clin. Microbiol. 1, 384-389 (1975)

/12/ Hayes, P.R.: J. Gen. Microbiol. 30, 1-9 (1963)

/13/ Herman, L.G. and Himmelsbach, C.K.: Hospitals 39, 72-76 (1965)

/14/ Herman, L.G.: Advance in Appl. Microbiol., 23, 155-171 (1978)

/15/ Hurwick, B.J.: Israel J. Med. Sci. 8, 602-605 (1972)

/16/ Karamian, N.A.: Am. Lab. 8, 24-28 (1976)

/17/ Kelsey, J.D., and Beeby, M.D.: The Lancet, July 11, 82-84 (1964)

/18/ King, E.D.: Am. J. Clin. Path. 31, 241-247 (1959)

/19/ Koyama U., Yazawa, Y., Yamogishi, S., and Arai T.: Jap. J. Microbiol. 18, 49-56 (1974)

/20/ McEntegart, M.C., and Clark, A.S.: Br. Dent. J. 134, 140-142 (1973)

/21/ Moffet, H.L., Allan D., and Williams, T.: Am. J. Dis. Child. 114, 13-20 (1967)

/22/ Olsen, H.: Danish Med. Bull. 14, 1-5 (1967)

/23/ Olsen, H.: Danish Med. Bull. 17, 171-172 (1970)

/24/ Olsen, H. and Ravin, T.: Acta. Path. Microbiol. Scand. Section B, 79, 102-106 (1971)

/25/ Otten, G. and Brown, G.D.: American Lab. 5, 49-60 (1973)

/26/ Pedersen, M.D., Marso, M.A., and Pickett, M.J.: Am. J. Clin. Path. 54, 178-192 (1970)

/27/ Pickett, M.J., and Pedersen, M.M.: Can. J. Microbiol. 16, 401-409 (1970)

/28/ Shewan, J.M.: J. Appl. Bact. 32, 1-4 (1969)

/29/ Spaepen, M.S., Bodman, H.A., Knudsin, R.B., Berryman, J.R., and Fencl, V.: Health Lab. Science 12, 316-320 (1975)

/30/ Stamm, W.E., Colello, J.J., Anderson, R.L., and Dixon, R.E.: New Engl. J. Med. 292, 1099-1102 (1975)

/31/ Stamm, J.M., Engelhard, W.E., and Parson, J.E.: Appl. Microbiol. 18, 376-386 (1969)

/32/ Starr, M.P., and Stephens, W.L.: J. Bact. 87, 293-302 (1964)

/33/ Weeks, O.B.: J. Appl. Bact. 32, 13-18 (1969)

/34/ Yabuuchi, E., Ohyama, A. Takeda, H., Sugiama, M., and Kono, S.: Jap. J. Microbiol. 14, 241-242 (1970)

# DISTRIBUTION OF CELLULOSE DEGRADING GRAM-NEGATIVE BACTERIA

Rosanne Mullings, J.H. Parish
University of Leeds, Department of Biochemistry, 9 Hyde Terrace,
Leeds LS2 9LS, U.K.

## Summary

New methods have been developed for the enrichment, purification and characterization of cellulose degrading bacteria from brackish, freshwater and terrestrial habitats. The methods involve the use of bacterial cellulose mats as cellulosic substrates for rapid identification and purification of these organisms. A preliminary survey of 53 isolates of mesophilic aerobic and facultatively anaerobic organisms reveals the following generalizations. (1) All the organisms were Gram-negative and no flagellates were found. (2) From any one habitat the cellulose degraders form tight clusters of related types. (3) The brackish samples from a salt marsh produced two novels types of myxobacteria, tentatively allocated to *Polyangium*. (4) From freshwater samples, the dominant types were cytophagas (5) An interesting group of organisms capable of degrading cellulose, agar and gelatin was discovered.

## Introduction

Degradation of cellulose occurs in several habitats and by several groups of microorganisms. Anaerobic fermentations based on cellulose are performed by specialized groups of bacteria in the gastrointestinal tracts of animals and notably in the rumen. Aerobic and facultatively anaerobic organisms are responsible for the degradation of cellulose in other environments. Wood-rotting and other cellulolytic fungi (belonging mainly to the ascomycete, imperfect and basidiomycete groups) are the characteristic cellulose degraders in conditions of low water activity. An exception is perhaps damp hay in which the dominant organism is *Thermoactinomyces* /2/. In water and soil the predominant cellulose degraders are bacteria. The occurrence of cellulose degradation among bacteria is seemingly sporadic and the literature is unsystematic; Gooday /4/ has written a recent review with comprehensive reference to earlier surveys.

Cellulose degradation is important for two reasons. (i) As cellulose is the most abundant form of fixed carbon in the biosphere, its breakdown is quantitatively an important component in the ecology of rotting vegetation and is also of practical importance in the treatment of sewage. (ii) The breakdown of cellulose to relatively reduced carbon compounds can, in principle, be achieved industrially. This is tantamount to interrupting the carbon cycle at one point so that, instead of being coupled directly to $CO_2$ formation, cellulose degradative processes can be harvested to yield fuels or carbon feedstocks for industry. In order to achieve this objective, the process can rely either upon the use of cell-free cellulase preparations or on microbial fermentations. In the former case the most appropriate source of enzymes may be the fungus, Trichoderma reesei; in the latter it is more appropriate to consider the use of a bacterium because of the greater ease of strain construction and genetic manipulation. In the light of this, we have devised methods for isolating and purifying cellulolytic bacteria. The methods were designed to identify only those organisms which can utilize cellulose as sole source of carbon and energy and which have no complex growth requirements. In this way we hope to have isolated organisms that are quantitatively important in their habitats as successful cellulose degraders. Our procedures eliminate relatively easily persistant contamination by adventitious heterotrophic bacteria that present large problems in conventional schemes for isolation of bacteria on paper as a cellulosic substrate. The results include characterization of 53 mesophilic cellulose degraders and allow preliminary conclusions to be drawn about the distribution of these organisms in certain habitats.

## Materials and Methods

Sources of samples: Blacktoft Sands is a brackish marsh (1.1% total salinity) colonized by Phragmitis reeds, in Humberside; Gledhow Valley samples were obtained from a pool produced in a dammed portion of a stream in a residential part of Leeds. Other samples were all collected in the Leeds area; the "aquarium" was a tropical fish tank.

Cellulosic substrates: paper used in these experiments was Whatman no. 1 filter paper; bacterial cellulose mats were produced from cultures of Acetobacter aceti xylinum by methods modified from Aschner /1/; wood pulp was prepared by de-resinifying sawdust and de-lignifying this with $NaClO_2/H^+$ and $Na_2SO_3$ by methods based on Green /5/. For certain experiments the wood pulp was fractionated to yield cellulose, hemicellulose A and hemicellulose B /13/. Analysis of fractions: dry weight yield of all fractions from sawdust was 71% (remainder presumed to be resins and lignin); the pulp was 54% cellulose, 22% hemicellulose A and 24% hemicellulose B; monosaccharides in acid hydrolysates of hemicellulose A were 48% xylose, 31% mannose and 21% glucose (by g.l.c. of trimethylsilyl derivatives); corresponding figures for hemicellulose B were 53% xylose, 33% mannose, 10% glucose and 4% glucuronic acid.

Stanier's nitrate medium /7/ was employed for most isolations; seawater medium consisted of 1.1% sea salts supplemented with nitrate and phosphate salts as in Stanier's nitrate. Enrichment media and isolation plates contained cycloheximide (25 $\mu g\ ml^{-1}$) and nystatin (5 $\mu g\ ml^{-1}$). All glucose media contained 1% glucose; the rich medium was Stanier's nitrate with glucose supplemented with peptone (0.5%) and yeast extract (0.5%). Agar was added to 1.5%; unless stated to the contrary all incubations were at room temperature.

Characterization: see ref. /9/ for the flagella stain and /3/ for other tests.

## Results

Organisms were isolated from several sources with and without enrichment using either paper or bacterial cellulose mats as substrates (Fig. 1). Similar or identical organisms were obtained from the same samples by each of the four methods; usually the highest yield and simplest purification was achieved by using enrichment followed by plating on paper.

In addition to organisms chosen for further study, we isolated two purple myxobacteria, tentatively identified as Polyangium, from the Gledhow samples. These appeared on paper and on a mat but did not degrade cellulose significantly. We also obtained a single colony (on paper) of a Gram-positive bacterium, Micromonospora bicolor, that degrades cellulose slowly. This organism probably arose by aerial contamination by a spore.

The properties of the majority of organisms are summarized in Table 1. None of these organisms turn pink in alkali and we conclude that the yellow and orange colours are not due to flexirubins /11/. Sori (fruiting bodies) were produced by strains of brackish origin only.

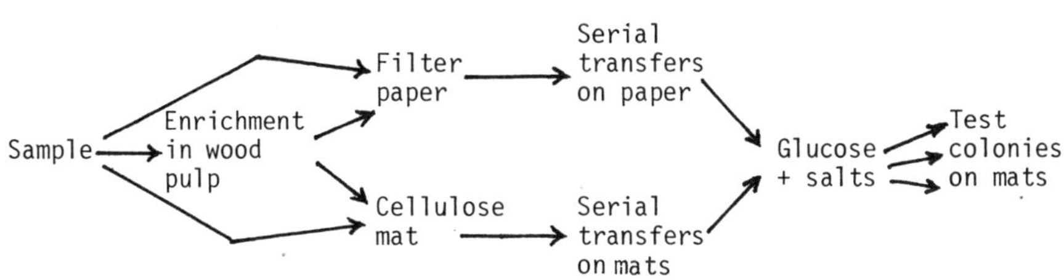


Fig. 1. Summary of methods for isolation and purification of cellulose degrading bacteria.

TABLE 1 (following two pages) Summary of properties of 48 cellulose degrading strains. A brackish water samples, B freshwater and terrestrial samples

---

Abbreviations used: + positive, - negative, ±unclear; untested or unconfirmed results are left blank. Other abbreviations are as follows.

| | |
|---|---|
| No. | RM strain number |
| From | Source of material; B Blacktoft, C compost, S strawberry soil, G Gledhow, D horse dung. |
| Purfn | Purification method; E enriched, N not enriched, P paper, M mats |
| Gram | Gram reaction |
| Cell | Shape of cells; r rods; sr slender rods (1.5-6 x 0.25-0.4 μm) |
| Col | Appearance of colonies on glucose salts: y, yellow becoming orange; g, gummy; s, spreading; f, formation of sori; p, pale |
| Diff | Diffusible pigment: S, only when deprived of seawater; R, in rich medium; V, variable, depending on salts mixture. |
| Sori | Presence or absence of sori: o, irregular, large and orange; b, irregular, small, deliquescent and brown. |
| Etch | Etching of agar by sori. |
| Mot | Motility: P, spreads over paper but non-motile under microscope; +P, motile both on paper and under microscope. |
| Fla | Flagella |
| Slime | Production of slime |
| Pad | Degradation of cellulose mat (pad). |
| Paper | Degradation of filter paper: T, becomes transparent, usually from edge; Y, yellow colour spreads over whole paper, degradation usually starts with darkening of colour at edge; O, colours entire paper, becoming orange; M, mottled mustard colour spreads from inoculum; B, mustard colour spreads over entire paper, degradation starts at edge, usually becoming brown. |
| PadG | Degradation of pad in the presence of glucose. |

Xyl (xylose), Gal (galactose), Lac (lactose), Suc (sucrose), Mtl (mannitol), Star (starch), Esc (esculin), Cello (cellobiose), HemiA and HemiB (hemicellulose A and B): growth on these as sole sources of carbon and energy.

| | |
|---|---|
| Rich | Growth in rich medium: S, same as Stanier's glucose; V, no sori; m, irregular mounds; f, greater tendency to form sori than on Stanier's glucose. |
| NoSal | Growth in absence of sea salts: f, greater tendency to form sori than in the presence of these. |
| NoNit | Growth in the absence of nitrate |
| H&L | Hugh and Leifson; I inert, O oxidative, F fermentative. |
| Ox | Oxidase |
| Cat | Catalase |
| Ure | Urease |
| 4°C and 42°C | Growth at these temperatures. |
| Res and Sens | Resistance and sensitivity to drugs (μg $ml^{-1}$); R rifampicin (5), C chloramphenicol (25), T tetracycline (10), A ampicillin (25), S streptomycin (15), N nalidixic acid (15), K kanamycin (25), Z sulphonilamide (10). Not shown in table: trimethoprim (25); RM27 was resistant; all other strains tested were sensitive. An entry in brackets for Res means that growth was impaired but not inhibited totally. |

TABLE 1.A BRACKISH WATER SAMPLES

| No. | 1 | 2 | 3 | 4 | 5 | 6 | 7 | 8 | 9 | 10 | 11 | 12 | 13 | 14 | 15 | 16 | 17 | 18 | 19 | 20 | 21 | 22 | 23 | 24 | 25 | 26 |
|---|---|---|---|---|---|---|---|---|---|---|---|---|---|---|---|---|---|---|---|---|---|---|---|---|---|---|
| From | BT | BT | BT | BT | BT | BT | BT | BT | BT | BT | BT | BT | BT | BT | BT | BT | BT | BT | BT | BT | BT | BT | BT | BT | BT | BT |
| Purfn | EM | EP | NM | NP | EM | NM | EM | EP | EP | NM | NM | NM | NM | NP | NP | EM | EM | EP | EP | EP | EM | EP | EP | EP | EP | EP |
| Gram | - | - | - | | - | - | - | - | - | - | | - | - | - | - | - | - | - | - | - | - | - | - | - | - | - |
| Cell | sr | sr | sr | | sr | r | sr | sr | sr | sr | | sr | sr | sr | sr | sr | sr | sr | sr | sr | sr | sr | sr | sr | sr | sr |
| Col | y | f | f | y | f | f | f | f | f | f | p | fs | | f | y | y | y | y | y | y | y | f | y | | y | f |
| Diff | +S | - | - | | +S | | +S | | | | | | | | | - | | +S | | - | +S | | +S | | | +S |
| Sori | +o | +b | +b | +o | +b | +b | | | +b | +b | | +b | +b | +b | +b | +o | +o | +o | +o | +o | +o | +b | +o | +o | +o | +b |
| Etch | | | | | + | | | | | | | | | | | | | | | | | | | | | |
| Mot | P | | | P | | | P | P | +P | P | | + | | P | | P | P | | P | P | P | P | P | P | P | P |
| Fla | | | | | | | | | | | | | | | | | | | | | | | | | | |
| Slime | + | | | | | | + | | + | | | + | | + | | + | + | | + | + | | | + | | | + |
| Pad | + | + | + | + | + | + | + | + | + | + | + | + | + | + | + | + | + | + | + | + | + | + | + | + | + | + |
| Paper | +Y | | | +0 | | | +Y | +Y | +B | | | +Y | | +Y | | +T | +Y | +0 | +0 | +Y (0) | +Y | +T | +0 | +0 | +0 | +B |
| PadG | + | | | + | | | - | - | - | | - | + | | - | - | + | + | | - | + | | - | | - | | - |
| Xyl | + | | + | + | | | - | + | + | + | + | | + | + | | + | + | + | + | + | + | + | + | | + | + |
| Gal | + | | + | + | | | - | + | + | | + | | + | + | | + | + | + | + | + | + | + | + | | + | + |
| Lac | + | | | + | | | - | + | + | | | | + | + | | ± | ± | - | + | + | + | + | + | | ± | + |
| Suc | ± | | | - | | | - | - | - | | | | ± | + | | - | - | - | - | ± | - | - | ± | | - | - |
| Mal | - | | | - | | | - | - | - | | | | ± | - | | - | - | - | - | - | - | - | - | | - | - |
| Star | +H | | | + | | | +H | +H | +H | + | + | | +H | +H | | - | +H | +H | +H | +H | +H | +H | +H | +H | +H | +H |
| Esc | - | | | | | | - | - | - | | - | - | | - | - | - | | | - | ± | | - | | - | | - |
| Cello | + | | + | + | | | + | + | + | + | + | | + | + | | + | + | + | + | + | + | + | + | | + | + |
| HemiA | + | | | | | | - | - | + | | - | - | | - | - | - | | | ± | ± | | - | | - | | - |
| HemiB | + | | + | | | | - | ± | + | | - | ± | | ± | - | ± | | + | ± | + | | - | | + | | - |
| Rich | S | | | | | S | SV | | | | | | | | | S | S | S | | | | | S | S | | |
| NoSal | f+ | | | | + | | f+ | + | f+ | | | + | | f+ | | + | + | f+ | + | f+ | f+ | + | + | f+ | | + |
| NoNit | - | | | - | | | - | - | - | | - | - | | - | - | - | | | - | - | | - | | - | | - |
| H&L | F | | | | | | I | I | I | | | | 0 | 0 | | I | I | I | I | Io | I | I | I | I | 0 | I |
| Ox | + | + | + | + | + | + | + | ± | + | + | - | + | - | - | + | + | + | - | + | + | + | - | + | + | ± | + |
| Cat | + | + | - | + | - | - | + | + | + | - | - | - | - | - | - | + | + | + | + | ± | + | + | + | + | - | + |
| Ure | - | | | + | | | - | - | + | | - | | | - | - | - | | | + | + | | - | | - | | - |
| 4°C | + | | | | | | - | - | + | | | | + | | | - | - | + | + | + | + | + | - | - | | - |
| 42°C | - | | | | | | - | - | - | | | | - | | | - | - | - | - | - | - | - | - | - | - | - |
| Res | | | | RT<br>AS<br>NK | AZ | | | | RC<br>TA<br>N | | | | | | | | | (S) | RT<br>AS<br>K | | | | | | | |
| Sens | RC<br>TA<br>SN<br>K | | | C | CT<br>SK | | RC<br>TA<br>SN<br>K | RC<br>TA<br>SN<br>K | SK | | | RC<br>TA<br>SN<br>K | | RC<br>TA<br>SN<br>KZ | RC<br>TA<br>SN<br>KZ | RC<br>TA<br>SN<br>KZ | RC<br>TA<br>SN<br>KZ | CT<br>A | CN | RC<br>TA<br>SN<br>KZ | | RC<br>TA<br>SN<br>KZ | | RC<br>TA<br>SN<br>KZ | | RC<br>TA<br>SN<br>KZ |

TABLE 1.B FRESHWATER AND TERRESTRIAL SAMPLES

| No. | 27 | 28 | 29 | 30 | 31 | 32 | 33 | 34 | 35 | 36 | 37 | 38 | 39 | 40 | 41 | 42 | 43 | 44 | 45 | 46 | 47 | 48 |
|---|---|---|---|---|---|---|---|---|---|---|---|---|---|---|---|---|---|---|---|---|---|---|
| From | SS | SS | C | C | C | C | C | C | C | C | C | C | C | C | C | C | C | C | G | G | G | D |
| Purfn | EP | NP | EP | EM | NM | NM | NM | NM | NM | NP | NP | NP | NP | NP | NP | NP | NP | NP | NP | NP | NP | NP |
| Gram | - | - | | - | - | - | - | - | - | - | - | - | - | - | - | - | - | - | - | - | | - |
| Cell | sr | sr | | r | sr | sr | sr | sr | sr | sr | sr | sr | sr | sr | sr | sr | sr | sr | sr | sr | | sr |
| Col | pg | yg | yg | yg | yg | yg | yg | yg | yg | yg | yg | yg | yg | yg | yg | yg | yg | yg | yg | y | yg | yg |
| Diff | - | +R | | | | - | - | - | - | - | - | - | - | - | - | - | - | - | - | - | - | +V |
| Sori | - | - | - | - | - | - | - | - | - | - | - | - | - | - | - | - | - | - | - | - | - | - |
| Mot | P | P | | | | | P | P | P | P | P | P | P | P | P | P | P | P | P | | | P |
| Fla | - | | | | | | | - | | | | | | | | | - | | - | | | - |
| Slime | ++ | + | + | | + | + | + | + | + | | | + | + | + | + | | + | + | + | | + | ++ |
| Pad | + | + | + | + | + | + | + | + | + | + | + | + | + | + | + | + | + | + | + | + | + | + |
| Paper | +T | +M | | | +Y | +Y | +Y | +Y | +Y | +Y | +Y | +Y | +Y | +Y | +Y | +Y | +Y | +Y | +Y | +Y | | +$\overset{Y}{M}$ |
| PadG | - | | | - | | - | - | - | - | - | - | - | | - | - | - | - | - | - | | | - |
| Xyl | + | - | | + | - | | ± | + | + | + | + | + | - | | - | - | - | - | - | | | + |
| Gal | | | | | - | | + | + | + | + | + | + | - | + | + | - | + | + | + | | | + |
| Lac | + | ± | | + | | + | + | + | ± | + | + | + | - | + | + | + | | + | + | | | + |
| Suc | - | - | | - | - | - | ± | ± | ± | ± | - | - | - | - | - | - | ± | ± | + | | | - |
| Mtl | - | - | | - | - | - | - | - | - | - | - | - | - | - | - | - | - | - | - | | | - |
| Star | +H | +H | | +H | | +H | +H | +H | +H | +H | +H | +H | - | +H | +H | - | +H | +H | +H | | | +H |
| Esc | - | | | | | | - | - | - | - | - | - | - | | - | - | - | - | - | | | - |
| Cello | + | | | + | | + | + | + | + | + | + | + | + | + | - | - | + | ± | + | | | |
| HemiA | - | | | | | | - | - | - | - | - | - | - | - | - | - | - | - | - | - | + | ± |
| HemiB | | | | | | | - | ± | - | ± | ± | ± | | - | - | + | - | + | + | - | + | + |
| Rich | Sm | Sf | | | S | S | SV | S | S | S | S | S | | | | S | S | | S | | | S |
| NoNit | ± | | - | | | | - | | ± | | - | - | - | - | - | | - | - | - | | - | ± |
| H&L | O | O | I | O | I | I | F | O | I | O | I | I | I | O | I | I | I | O | O | | O | O |
| Ox | + | + | | + | + | + | + | + | + | + | + | + | - | + | + | + | | - | + | | + | + |
| Cat | + | + | | + | ± | ± | ± | ± | ± | + | + | ± | + | + | ± | + | | - | + | | ± | + |
| Ure | - | | | | | | - | - | - | - | - | - | - | - | - | - | - | - | - | | | - |
| 4°C | - | - | - | | - | - | + | + | + | + | + | + | - | + | + | - | + | + | + | - | - | + |
| 42°C | - | - | - | | - | - | - | - | - | - | - | - | ± | - | - | - | + | - | - | - | - | - |
| Res | AZ | | | | | | AZ | A | AZ | AZ | AZ | A | | (A) | (Z) | AZ | | A (S) | AZ | | | A |
| Sens | RC TS NK | | | | | | RC TS NK | RC TS NK | RC TS NK | RC TS NK | RC TS NK | RC TS Z | | RC TS Z | RC TA K | RC TS | RC TA K | RC TN KZ | RC TS NK | | | RC TS Z |

We also obtained some organisms capable of degrading both cellulose and agar and also gelatin (Table 2). These organisms have been less extensively studied so far. They were obtained from several sites including a different salt marsh in the South of England. Superficially these organisms appear to be similar; they too lack flexirubins.

Table 2. Bacteria capable of degrading both cellulose and agar

| Strain no. | RM49 | RM50 | RM51 | RM52 | RM53 |
|---|---|---|---|---|---|
| Origin (a) | AS | GS | M | RN | RB |
| Growth on Cellulose (b) | + | + | + | + | + |
| Effect of Glucose (c) | N | N | N | P | P |
| Growth on Hemicellulose A | - | - | - | - | - |
| Growth on Hemicellulose B | - | ± | ± | ± | - |
| Growth on agar (b) | + | + | + | + | + |
| Growth on gelatin | + | - | + | + | - |
| Liquefaction of gelatin | - | - | + | - | - |
| Gram reaction | - | - | - | | |
| Dimensions (μm) | 2.3 x 0.3 | | 2-4 x 0.5 | | |

Notes (a) AS aquarium sediment, GS garden soil, M marsh near River Colne (Essex), RB mud on banks of this river.

(b) as sole source of carbon and energy.

(c) N no effect on cellulose degradation, P partial inhibition of this.

## Discussion

We believe that of the methods described, the simplest and most reliable method for obtaining cellulose degrading bacteria is enrichment in wood pulp medium followed by plating and serial transfer on paper (about four passages), plating on glucose and testing on cellulose mats.

From the present survey we conclude that the dominant cellulose degrading bacteria in aquatic habitats belong to the myxobacteria or the cytophaga group. Our tentative assignments are as follows.

The strains from Blacktofts Sands appear to fall into two groups: those with small brown sori (such as RM12) and those with larger orange sori (such as RM16). From the appearance of the cells and sori, the organisms correspond to members of the genus, Polyangium /10/. However, they differ from P. cellulosum strains in several respects: the pattern of paper degradation is quite different and the sori are never produced on cellulosic substrates. We believe that with certain exceptions, such as RM7, all the Blacktoft strains belong to a fairly homogeneous group of salt tolerant, cellulose degrading myxobacteria, probably polyangia; the majority are oxidase positive and will grow on cellulose, glucose, cellobiose, xylose, galactose, lactose and starch; catalase and the Hugh and Leifson reactions are variable characters in this group.

The majority of the freshwater and terrestrial strains are seemingly related. From the published taxonomy /8/, the organisms correspond to Cytophaga hutchinsonii. In the light of current revisions of the status of the genus /12/, this allocation must remain tentative. It is noteworthy that none of these organisms appear to contain flexirubins. No cellulose degrading, cyst forming organisms were found in this survey. Moreover, we are confident that these organisms (such as RM40 and RM45) are not acystogenous variants of Sporocytophaga myxococcoides for two reasons: they do not form the yellow "stained holes" in paper that are typical of Sporocytophaga and the latter organism demonstrates a good "flexirubin reaction" when alkali is added to colonies. Strains RM35 and RM 48 grow weakly without nitrate and may be capable of fixing $N_2$.

The cellulose-and-agar degrading strains (RM49-RM53) also appear to be cytophagas lacking flexirubins. The only species listed as being able to degrade both cellulose and agar, C. krzemienskae and C. diffluens, are reported as requiring salt /8/. Although strains RM51, RM52 and RM53 were from a saline environment, RM49 and RM50 were not and none of these strains require salt for growth.

We are currently involved in completing the characterization of the strains, surveying a wider range of habitats and examining the regulation of polysaccharide degradation by the cellulolytic and agarolytic organisms such as Cytophaga sp. RM51.

## Acknowledgements

We are very grateful to Dr. M.F. Chaplin for advice and help with the gas-liquid chromatography of sugar derivatives and Mr. A. Grieve, warden of the RSPB bird sanctuary at Blacktoft Sands, for help and access to samples. R.M. is supported by a NERC Research Studentship.

## References

/1/ Aschner, M.: J. Bacteriol. 33, 249-252 (1937)

/2/ Bellamy, W.D.: Development in Industrial Microbiology 18, 249-253 (1977)

/3/ Cruickshank, R., Duguid, J.P., Marmion, B.P., Swain, R.H.A.: Medical Microbiology, 12th edition, vol. 2. Edinburgh (1975): Churchill Livingstone

/4/ Gooday, G.W.: p. 437-460 in: Berkeley, R.C.S., Gooday, G.W., Ellwood, D.C. (eds.), Microbial polysaccharides and polysaccharases. London (1979): Academic Press

/5/ Green, R.W.: p. 9-21 in: Whistler, R.L. (ed.), Methods in Carbohydrate Chemistry, vol. III. New York (1963): Academic Press

/6/ v. Hofsten, B., Berg, B., Beskow, S.: Arch. Microbiol. 79, 69-79 (1971)

/7/ Lampky, J.R.: J. Bacteriol. 126, 1278-1284 (1976)

/8/ Leadbetter, E.R.: p. 101-112 in: Buchanan, R.E., Gibbons, N.E. (eds.), Bergey's Manual of Determinative Bacteriology, 8th edition. Baltimore (1974): Williams & Wilkins

/9/ Mayfield, C.I., Inniss, W.E.: Can. J. Microbiol. 23, 1311-1313 (1977)

/10/ McCurdy, H.D., Peterson, J.E.: p. 92-96 in: Buchanan, R.E., Gibbons, N.E. (eds.), Bergey's Manual of Determinative Bacteriology, 8th edition. Baltimore (1974): Williams & Wilkins

/11/ Reichenbach, H., Achenbach, H., Kohl, W.: this volume

/12/ Reichenbach, H., Behrens, H., Hirsch, I.: this volume

/13/ Whistler, R.L., Feather, M.S.: p. 144-145 in: Whistler, R.L. (ed.), Methods in Carbohydrate Chemistry, vol. V. New York (1965): Academic Press

# SOME ECOLOGICAL ASPECTS OF AQUATIC CYTOPHAGAS

**W. Reichardt**

Department of Microbiology
Oregon State University
Corvallis, Oregon 97331, USA

## Summary

*Cytophaga*-like bacteria are at least temporarily abundant in almost all aquatic habitats, although the techniques which are currently available for large scale identification and enumeration, may only lead to rather vague taxonomic conclusions. As for obtaining maximal plate counts, the enumeration of relatively fastidious and mostly organic nitrogen-requiring marine forms could be considerably improved by using 38 mg/l of polymyxin B as a counterselective agent.

From a collection of chitin-degrading and predominantly estuarine bacteria, Cytophagaceae-strains were selected to study utilization of glucose, production of depolymerizing enzymes and survival under hydrostatic pressure. While swarming (spreading) as one of the most striking colony features was insufficient as taxonomic indicator, *Cytophaga*-like isolates could be classified on the basis of their substrate requirements for swarming. Different swarming types could at least partly be ranked according to their kinetic parameters for glucose utilization. While being able to degrade various biopolymers effectively even at low water temperatures, psychrotrophic, oligonitrophilic isolates suggest a key position for this type of freshwater cytophagas in major biogeochemical pathways of the carbon cycle. Growth rates, production of chitin-degrading enzymes as well as autolytic activities of one of these isolates (*Cytophaga johnsonae*, strain C 21) were characterized by similar temperature responses. In contrast to deep freshwater habitats, nothing is known about a corresponding contribution of marine *Cytophaga* spp. to biodegradation processes in the deep sea. It could be shown, however, that even a starved culture of an estuarine surface water isolate was able to survive hydrostatic pressures up to 500 at .

## Introduction

Distribution patterns which show an occasional predominance of Cytophaga-like bacteria in various aquatic habitats /f.e. 26,28,38,39,40,42/ seem to indicate an opportunistic strategy of population dynamics /45/. But considering the possible impact of these organisms on biogeochemical nutrient cycles, only little is known about single environmental factors which may control their abundance and survival /40, 42,44,47/.

The ecological aspects covered in this brief survey refer to the influence of organic substrates, temperature, and hydrostatic pressure on growth and enzymatic activities. Most recent data included in this report were obtained with Cytophagaceae strains which had been isolated for the purpose of a numerical taxonomy study of chitin-degrading bacteria from predominantly estuarine waters.

As culture experiments are necessary, but not sufficient to understand the ecology of aquatic cytophagas, improved enumeration techniques for field studies have to be developed. As a first attempt in this direction, a counterselective plate count technique for the enumeration of marine Cytophaga-like bacteria will be described.

## Materials and Methods

Marine water samples were collected at the Oregon Pacific coast near Newport. Strains for culture experiments were selected from a collection of 70 chitin-degrading Cytophaga-like isolates from estuarine, marine, and freshwater habitats (Chesapeake Bay, Oregon Pacific coast, and lakes in Oregon, California and West Germany).

Media for marine and estuarine bacteria were prepared using dilute, 0.45 µm -membrane-filtered seawater (FSW), or an artificial seawater substitute (ASW) containing 24 g/l NaCl, 5.3 g/l $MgCl_2 \cdot 6 H_2O$, 7.0 g/l $MgSO_4 \cdot 7 H_2O$, and 0.7 g/l KCl in distilled water. Colony forming units (CFU) of marine Cytophaga-like bacteria were enumerated on the following media: Seawater agar (SW) with 0.2 mM $KH_2PO_4$-$Na_2HPO_4$-buffer and 15 g/l Difco agar in 3/4 strength FSW; SW-PB consisting of SW and 38 mg/l of polymyxin B sulphate (Sigma Chemicals); SW-PEN consisting of SW and 10 mg/l of penicillin G (Nutritional Chem.Corp.); TSB consisting of SW and 8 g/l trypticase soy broth (BBL); TSB-PB containing 38 mg/l polymyxin B in TSB; and TSB-PEN containing 10 mg/l penicillin G in TSB. Using a spread plate technique, samples were diluted in 1/3 strength ASW containing 0.2 mM $KH_2PO_4$-$Na_2HPO_4$-buffer and 10 mg/l tween 80 and incubated for 14 d at 22°C. The preliminary routine identification of marine Cytophaga-like bacteria was essentially based on cellular morphology while screening representative colonies under a phase contrast microscope.

Utilization of glucose was determined by measuring the rates of both 'incorporation' into the macromolecular,10% cold TCA-precipitate- fraction, and $CO_2$-evolution ('respiration') during incubation with $^{14}C$ (U)- glucose (S.A.= 284 mCi/mM, Amersham) /3/. Cells grown in membrane filter-sterilized liquid media containing 1/3 strength ASW, basal mineral medium ( 0.1 mM $K_2HPO_4$-$Na_2HPO_4$-buffer, 1.0 g/l $(NH_4)_2SO_4$, 0.2 g/l $MgSO_4 \cdot 7H_2O$, 3 mg/l $CaCl_2$ and 0.3 mg/l $FeCl_3$), 10 mM glucose and,if required,8g/l Casamino acids (Difco), were harvested during their stationary growth phase, centrifuged twice (7000 g) and immediately resuspended in 1/3 str.ASW- mineral medium to start the assay. 9 ml aliquots of the suspension (adjusted to an optical density of 0.010 (550 nm,1cm) were incubated for 30 min at 22°C in 50 ml-flasks with 0.1 μCi $^{14}C$ (U)-glucose diluted with 5,10,20,40,and 80 μMol/l of unlabelled glucose (rotary shaker at 100 rpm).After acidification and $CO_2$-absorption using phenylethylamine, incorporation and respiration of glucose were measured using a Beckman LS 100C liquid scintillation counter / 3, 17, 20/.If applicable,the kinetic parameters of $V_{max}$ and $k_M$ were calculated using the linear transformation of the Michaelis-Menten equation according to Eadie /10/.

Autolytic activities of cell suspensions were determined after exponential growth in a polythermostat incubator (Scientific Industries Inc.,Mineola,N.Y.), centrifugation (15 min, 6000 g), resuspension in basal mineral medium (see above) and re-incubation at the same temperature by recording the decrease of absorbance (450 nm) in the side-arm culture tubes using a Bausch & Lomb Spectronic 20 photometer.

For hydrostatic pressure experiments, cell suspensions in 1/3 strength ASW- mineral medium (absorbance at 550 nm, 1cm: 0.010) were loaded into 3 ml disposable syringes capped with sealed hypodermic needles and put into special pressure cylinders at 5°C / 34, 36/.Plate counts of appropriate dilutions were determined on Difco plate count agar in 1/3 strength ASW after 4 d incubation at 22°C.

## Results and Discussion

### OCCURRENCE

Due to an apparent lack of rapid routine techniques still being accurate enough to identify and enumerate members of the genus *Cytophaga* on a large scale, reports about population densities of aquatic cytophagas represent only rough estimates of *Cytophaga*-like organisms.

When using selective enumeration techniques which are mainly based on colonial and cellular morphology and, in certain cases, also selective or counterselective additions such as chitin **and cellulose** or polymyxin B, there seems to be hardly one aquatic habitat in which Cytophaga -like organisms ( presumptive cytophagas) could not at least temporarily be detected as one of the major groups of the heterotrophic bacteria being determinable on ordinary plate count media. Maximal counts of presumptive cytophagas may even exceed 80 % of the total colony forming units (CFU s) on casein-pepton-starch-glycerol-agar (CPS) /9/, which is often used for plate counts of freshwater bacteria. (Table 1).

The percentage stated in Table 1 for marine cytophagas on non-selective media might appear relatively low. It must be mentioned, however, that in aged seawater with a considerably lower content of degradable organic matter and accordingly lower CFU counts, Cytophaga-like bacteria accounted for approximately 30 % of the total CFU count (see Table 2). While this could already indicate a preference of aquatic cytophagas for oligotrophic habitats, such an influence must not be overestimated. As shown in Table 1, extremely high numbers of presumptive cytophagas have been reported not only for natural freshwater /7, 8, 12, 13, 22, 28, 39, 40, 41, 42, 44, 45, 49 / or marine habitats / 25, 26, 30, 32, 38 /, but also such extreme environments as sewage / 4, 14, 29/ and drinking water / 11, 19, 59, 62, Reichardt & Mueller, unpubl. report/. Moreover, aquatic Cytophaga spp. belong to the predominant microflora which colonizes and decomposes plant litter / 31, 39, 54/, and have also been reported as one of the principal constituents of the epizoic bacterioflora associated with fishes / 21, 53, 57/.

Table 1. Maximal CFU counts reported for presumptive aquatic cytophagas

| Habitat, Reference // | Plate count medium | Identification | CFU/ml or mg | % of total, on: |
|---|---|---|---|---|
| FRESHWATER, creek /7/ | pep.milk agar | Col.color, cell morphology | $7 \cdot 10^4$ | 85, pep.milk A. |
| eutroph.lake/40/ | CPS agar | Col.color, cell morphol., spherobl. | $2.5 \cdot 10^5$ | 81, CPS A. |
| activated sludge/4/ | Sewage agar | Col.color, cellulose-degrading | $6.2 \cdot 10^7$ +) | 12, Sewage A. |
| lake sediment /45/ | CPS agar | Orange spreading col., flexirubin+ | $2.6 \cdot 10^4$ | 3, CPS A. |
| drinking water/11/ | Mouton agar | Col.spreading, translucent, iridesc. | $10^3$ | 100 Mouton A. |
| ESTUARINE, polluted station, Chesapeake Bay | Chitin -PCA *) | yellow-orange, spreading, flexirubin+ chitindegrading | $4.4 \cdot 10^3$ | 34 Chitin-PCA |
| MARINE, upwelling area, NW Pacific Coast | SW-PB agar **) | Colony color, Cell morphol. | $1.7 \cdot 10^4$ | 0.8 SW Agar |

+) calculated from reported results. *) **) Reichardt, unpublished, see also Table 2
Chitin-PCA = Difco- plate count agar containing 5 g/1 colloidal chitin in 1/3 str.ASW.

ENUMERATION OF MARINE CYTOPHAGAS

Most Cytophaga plate count media are designed for freshwater samples / 5, 7, 24, 43, 63/. On the other hand, an agar medium used for cellulose decomposing marine Cytophaga / 25/ might be too selective considering f.e. the widespread requirement for organic nitrogen among marine and estuarine types /48, 50/. A numerical taxonomy study (Reichardt & Colwell, unpubl.) finally led to the conclusion that Cytophaga-like bacteria from non-freshwater environments may be generally more resistant to polymyxin B than freshwater isolates. Moreover, this antibiotic had already been recommended for the enrichment of 'cryptic' marine chromogenic bacteria /37/. As shown in Table 2, the addition of 38 mg/l polymyxin B sulphate to marine plate count media led to a striking increase of Cytophaga-like colony forming units, partly by a factor of 100.

Table 2. Comparison of non-selective *), selective **) and counterselective ***) plate count media for marine Cytophaga-like bacteria

| Type of medium: | SW *) | SW-PB ***) | SW-PEN ***) | TSB **) | TSB-PB ***) | TSB-PEN ***) |
|---|---|---|---|---|---|---|
| Sample 1: Decaying Macrophytes | | | | | | |
| A) Total CFU/ml ($x\ 10^6$) | 560 | 138 | 500 | 230 | 5 | 13 |
| B) % **Cytophaga-like** CFU | 0.2 | - | 0.4 | - | 100 | - |
| C) % spreading colonies +) | 100 | - | 100 | - | 100 | - |
| Sample 2: Normal Seawater | | | | | | |
| A) Total CFU/ml ($x\ 10^4$) | 220 | 6 | 200 | 180 | 0.1 | 17 |
| B) % Cytophaga-like CFU | - | 30 | - | - | 90.5 | - |
| C) % spreading colonies +) | - | 10.7 | - | - | 0 | - |
| Sample 3: Seawater, aged 6 months | | | | | | |
| A) Total CFU/ml ($x\ 10^4$) | 207 | 72 | 149 | 77 | 63 | 74 |
| B) % Cytophaga-like CFU | 8.2 | 22.2 | 17.4 | 31.2 | 100 | 32.4 |
| C) % spreading colonies +) | 100 | 56.3 | 73.6 | 33.3 | 38.1 | 87.7 |

Average of triplicate plate count determinations. +) refers to fractions of B).

Obviously, the counterselective effect of polymyxin B depended on the concentration of organic matter in the samples as well as in the agar media. Enrichment from marine samples with an extremely high organic matter content and accordingly high concentrations of heterotrophic bacteria (sample 1) was only successful on nutrient-rich polymyxin B-agar. Nutrient-poor 'aged' seawater, on the other hand, (sample 3) allowed a substantial colony growth of Cytophaga-like bacteria, even if polymyxin B had been omitted. The counts obtained for total heterotrophic and Cytophaga-like bacteria in samples 1 and 3 could indicate that during aging of seawater Cytophaga populations were able to enhance their relative abundance by at least a factor of 20.

The widespread resistance to penicillin G which has been demonstrated for freshwater cytophagas and applied for their selective enumeration / 29, 63/, could not be verified for marine isolates. Neither proved 10 mg/l penicillin G useful as a counterselective addition to plate count media for marine Cytophaga-like bacteria . Nevertheless, this antibiotic accounted for a higher percentage of spreading Cytophaga-like colonies in sample 3.

## SWARMING AND UTILIZATION OF DISSOLVED ORGANIC SUBSTRATES

The spreading (swarming) habit of Cytophaga -like colonies belongs to the key features which are often used in simplified procedures to identify and enumerate these bacteria (f.e./7, 12, 13/). Since swarming has long been known to depend often on low nutrient and agar concentrations /50/, counts of swarming colonies on ordinary plate count media may considerably underestimate the real number of Cytophaga colonies. On the other hand, the alternative use of nutrient-poor media would most likely exclude certain predominantly marine cytophagas which are nutritionally most fastidious, poor swarmers (Reichardt,unpubl.). Moreover,multiple transfer experiments with an estuarine isolate have indicated that even cloned cultures can split by an unknown mechanism into swarming and non-swarming descendants (Reichardt,in prep.). All these observations lead to the conclusion that colonial swarming is an extremely unstable and sometimes even unpredictable property,which can hardly be sufficient as 'key indicator' for enumeration of Cytophaga-like colonies.

Another ecologically important aspect leads to the question,if and how different swarming habits could possibly reflect different patterns of the utilization of organic carbon. Estuarine and marine Cytophaga-like strains showed at least four different swarming responses depending on the agar media used (Table 3).

If swarming is affected by different types and concentrations of organic substrates, also kinetics of organic carbon utilization are likely to differ among the strains assigned to different 'swarming categories'.A few data obtained for glucose incorporation of a limited number of strains suggest indeed that such a relationship might really exist (Table 3). Strain E-4, which was characterized by its utilization of carbohydrates as sole carbon sources and carbohydrate-dependent formation of swarming colonies ( category : sw (ch) ),showed maximal glucose incorporation rates. For comparison, a marine isolate (M-134) representing the other extreme as a nutritionally more fastidious poor swarmer (category: sw -) ,was characterized by extremely low glucose incorporation ($V_{max}$) . At the same time, the maximal value obtained for the Michaelis-Menten constant ($k_M$) of glucose respiration indicates a very low substrate affinity. Yet, there was one exception from this trend, since strain E-78, a flexirubin positive, salt requiring estuarine member of the sw- —category, was able to incorporate glucose at a rather high rate ($V_{max}$). In contrast to the other strains, however, minimal $k_M$-values for both incorporation and respiration indicated that this isolate had an extremely high affinity for glucose.

Obviously such a kinetic approach to study relationships between substrate availability and swarming could reveal certain possible trends which might be helpful for further mechanistic explanations. Yet, it still remains unsatisfactory, as long as complementary data about the transport systems involved and basic information about chemotactic responses of these bacteria are not available.

Table 3. Comparison of swarming types and kinetic parameters of glucose utilization determined for estuarine and marine *Cytophaga*-like isolates

| Strain # | Type of Swarming *) | Glucose as sole C-source | Utilization of Glucose: Incorporation $V_{max}$ +) | Incorporation $k_M$ +) | Respiration $V_{max}$ +) | Respiration $k_M$ +) |
|---|---|---|---|---|---|---|
| E-118 | sw + | − | − **) | − **) | 23.9 | 22 |
| M-134 | sw - | − | 2.1 | 15 | 23.7 | 645 |
| E-8 | sw (pep) | + | 5.2 | 22 | 9.5 | 29 |
| E-9 | sw + | + | 10.9 | 58 | 8.7 | 37 |
| E-78 | sw - | − | 16.4 | 6 | 27.7 | 3 |
| E-4 | sw (ch) | + | 38.0 | 31 | 16.2 | 12 |

Strains are ranked according to their $V_{max}$ for glucose incorporation.

*) Swarming types were assigned to the following categories:

sw - : formation of swarming colonies restricted under most growth conditions;
sw (pep) : swarming colonies formed preferably on agar media containing peptone;
sw (ch) : swarming colonies formed preferably on agar media containing carbohydrates as sole carbon source;
sw + : formation of swarming colonies largely independent of organic carbon source.

**) Michaelis-Menten kinetics not applicable. +) $V_{max}$: [$10^{-9}$ nM·min$^{-1}$·cfu$^{-1}$]; $k_M$: [µM/l].
Plate counts (CFU) determined on Difco-plate count agar containing 1/3 strength ASW.

## DEPOLYMERIZING ENZYMES

The ecologically most remarkable organic substrates for aquatic cytophagas are biopolymers. Due to their capacity to depolymerize a large variety of proteins /6, 23, 32, 52/ or polysaccharides / 1,2, 6, 15, 16, 18, 42, 55, 56, 58, 60/, aquatic *Cytophaga* spp. fill a key position in the biogeochemical pathways of the carbon cycle.

Psychrophilic properties, which are at least known for enzymes involved in the degradation of protein /32/ and chitin /42, 43, 46/, indicate important adaptations to the low-temperature environment of deeper water layers and sediments, where the final steps of the biodegradation of aquatic detritus and seston are most likely to occur. Part of these depolymerizing enzymes might also account for lytic activities which were found most frequently in freshwater cytophagas, but are also known for marine isolates (Reichardt, unpubl./33/).

An example for autolytic activities of a freshwater isolate is shown in Fig.1. Rates of autolysis showed a pattern of temperature dependence similar to that of the growth rates. Also, a certain coincidence with the double-peaked curves obtained earlier for the temperature dependence of chitin-degrading enzymes /43, 46/ was observed.

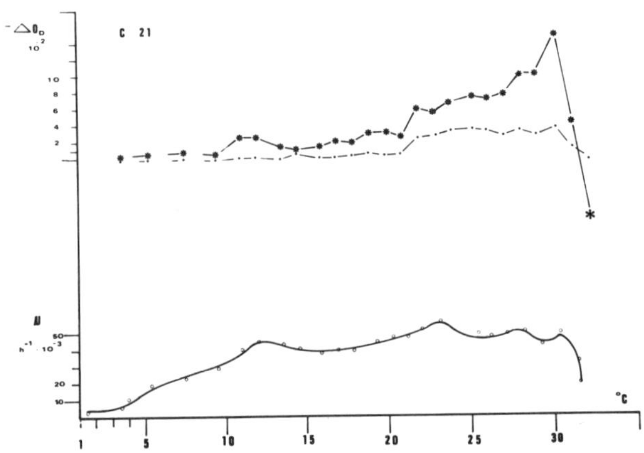


Fig.1. Influence of temperature on exponential growth rates ( $\mu$) and starvation-induced autolytic activities ( $-\Delta O_D$) of freshwater isolate C 21 (Cytophaga johnsonae). Decrease of absorbance ( $-\Delta O_D$) was determined 80 min (point symbols) and 14 hrs (star symbols) after re- incubation at same temperature. Growth in: Basal mineral salts medium /43/+ 10 mM N-acetylglucosamine; Re-incubation in: Basal mineral salts medium only.

If autolytic enzymes play a role by providing endogenous nutrients for a smaller part of a starving population, is not known yet. However, it seems remarkable that a steep decrease of absorbance (turbidity) of starved, re-incubated cell suspensions was obtained at near-maximum growth temperatures. Apparently this was associated with the formation of star shaped aggregates at 31 °C ( big star symbol). The influence of this phenomenon on the accuracy of absorbance values as a measure of lytic activities was not determined.

## SURVIVAL UNDER HYDROSTATIC PRESSURE

Deep and cold freshwater habitats which may be depleted of organic nitrogen substrates, but store large quantities of sestonic polysaccharides, appear as an ideal ecological niche for the oligonitrophilic and psychrotrophic group of freshwater cytophagas / 42, 43, 44, 45, 48/. If, however, deep sea habitats as the corresponding ultimate destination for degradable biopolymers in the marine environment, can likewise ensure the survival of marine cytophagas, has still to be confirmed.

The most characteristic ecological factor in the deep sea being not encountered in freshwater environments is extreme hydrostatic pressure (up to 1000 at). Therefore it seems worth mentioning that even a salt-requiring Cytophaga isolate from estuarine surface water was able to survive well under moderate hydrostatic pressure being equivalent to 2000 - 5000 m water depth. Survival at 100 and 200 atm was even better than

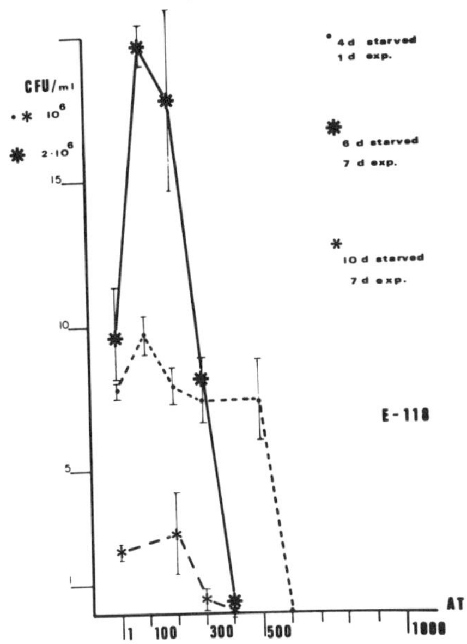


Fig.2. Influence of hydrostatic pressure (At) on survival rates of estuarine Cytophaga isolate E-118 at 5°C after 4,6,and 10 days of starvation in mineralmedium/43/ containing 1/3 strength ASW.

at 1 atm. Extended starvation caused an increase of barotolerance in agreement with earlier reports about a psychrophilic Vibrio /34/.

Since the first truly barophilic bacteria have only recently been cultivated, (A.Yayanos, Scripps Inst.of Oceanography, personal communication), any speculation about barophilic properties of marine cytophagas would be premature at the moment.

## References

/1/ Agbo,J.A.C & Moss,M.O.: J.Gen.Microbiol.115, 355-368 (1979)

/2/Bacon,J.S.D., Gordon,A.H., Jones,D., Taylor,I.E., Webley, D.M.: Biochem.J.120, 67-78 (1970)

/3/ Baross,J.A., Hanus,F.J., Griffith,R.P., Morita,R.Y.: J.Fish.Res.Bd.Can. 32, 1876-1879 (1975)

/4/ Benedict,R.G., & Carlson,D.A.: Water Research 5, 1023 - 1030 (1971)

/5/ Carlson,R.V., & Pacha,R.E.: Appl.Microbiol.16, 795 - 796 (1968)

/6/ Chattaway,F.W., Shenolikar,S., O'Reilly,J., Barlow,A.J.E.: J.Gen.Microbiol. 95, 335 - 347 (1976)

/7/ Christensen,P.J., & Cook,F.D.: Can.J.Microbiol. 18, 1933 - 1940 (1972)

/8/ Christensen,P.J.: Arctic 21, 309 - 311 (1974)

/9/ Collins,V.G., & Willoughby,L.G.: Arch.Mikrobiol. 43, 296 - 307 (1962)

/10/ Eadie, G.S.: J.Biol.Chem. 146, 85 - 93 (1942)

/11/ Graef,W., & Pelka,G.: Zbl.Bakt.Hyg.I Orig.B 169, 225 - 239 (1979)

/12/ Graef,W., & Stuerzenhofecker,P.: Arch.Hygiene 148, 79 - 96 (1964)

/13/ Graef,W., & Stuerzenhofecker,P.: Arch.Hygiene 149, 256 - 265 (1965)

/14/ Guede,H.: Appl.Environ.Microbiol. 39, 756 763 (1980)

/15/ Guede,H., & Reichardt,W.: Water Research 7, 1881 - 1883 (1973)

/16/ Gunja-Smith,Z., Marshall,J.J., Smith,E.E., Whelan,W.J.: FEBS Letters 12,96 - 100 (1970)

/17/ Harrison,M.J., Wright,R.T., Morita,R.Y.: Appl.Microbiol. 21, 698 - 702 (1971)

/18/ Hedges,A., & Wolfe,R.S.: J.Bact. 120, 844 - 853 (1974)

/19/ Herman,L.G.: Health Lab.Sci. 13, 5 - 10 (1976)

/20/ Hobbie,J.E., Crawford,C.C.: Limnology & Oceanography 14,528 -532 (1969)

/21/ Horsley,R.W.: J.Appl.Bact. 36, 377 - 386 (1973)

/22/ Horsley,R.W.: J.Appl.Bact. 46, 507 - 520 (1979)

/23/ Jackson,R.C., Wolfe,R.S.: J.Biol.Chem. 243, 879 - 888 (1968)

/24/ Jeffers,E.: Intern.Bull.Bacteriol.Nomencl.Taxon. 14, 115 - 136 (1964)

/25/ Kadota,H.: Bull.Jap.Soc.Sci.Fish. 19, 476 - 480 (1953)

/26/ Kadota,H.: Bull.Jap.Soc.Sci.Fish. 20, 120 - 124 (1954)

/27/ Kelch,W.J., & Lee,J.S.: Appl.Environ.Microbiol. 36, 450 - 456 (1978)

/28/ Lighthart,B.: Can.J.Microbiol. 21, 392 - 394 (1975)

/29/ Lighthart,B., & Loew,G.A.: J.Water Poll.Contr.Fed. 44, 2078 - 2085 (1972)

/30/ Lovelace,T.E., Tubiash,H., Colwell,R.R.: Bact.Proc. 1967, 46 (1967)

/31/ Mason,C.F., & Bryant,R.J.: J.Ecol. 63, 71 - 96 (1975)

/32/ McDonald,I.J., Quadling,C., Chambers,A.K.: Can.J.Microbiol. 9, 303 - 315 (1963)

/33/ Miyamoto,S., Kuroda,K., Hanaoka,M., Okada,Y.: Jap.J.Microbiol. 20, 517 -527 (1976)

/34/ Novitsky,J.A., & Morita,R.Y.: Appl.Environ.Microbiol. 32, 617 - 622 (1976)

/35/ Novitsky,J.A., & Morita,R.Y.: Marine Biology 49, 7 -10 (1978)

/36/ Oppenheimer,C.H., & ZoBell, C.E.: J.Mar.Res. 11, 10 - 18 (1952)

/37/ Pratt,D., & Reynolds,J.: p.258 - 267 in: Colwell,R.R. & Morita,R.Y. (eds.), Effect of the ocean environment on microbial activities. Baltimore(1974) : University Park Press

/38/ Quadling,C., & Colwell, R.R.: Bact.Proc. 1963, 40 (1963)

/39/ Ramsay,A.J.: New Zealand J.Mar.Freshw.Res. 11, 541 557 (1977)

/40/ Reichardt,W.: Verh.Internat.Verein.Limnol.(Leningrad) 18, 1277 - 1286 (1973)

/41/ Reichardt,W.: Zbl.Bakt.I Ref.233, 33 - 34 (1973)

/42/ Reichardt,W.: Zbl.Bakt.Hyg.I Orig.A 227, 85 - 93 (1974)

/43/ Reichardt,W.: Verh.Internat.Verein.Limnol. 19, 2636 - 2642 (1975)

/44/ Reichardt,W.: Verh.Internat.Verein.Limnol. 20, 2227 - 2233 (1978)

/45/ Reichardt,W.: Swiss J.Hydrol. 40, 249 - 261 (1978)

/46/ Reichardt,W.: ASM Annual Meeting Abstracts, N 79 (1979)

/47/ Reichardt,W.: ASLO Annual Winter Meeting,Abstracts (1980)

/48/ Reichardt,W., Morita,R.Y., Colwell,R.R.: ASM Annual Meeting Abstracts,**N63(1980)**

/49/ Ruschke,R., & Rath,M.: Arch.Hydrobiol./ Suppl. 28, 377 - 402 (1966)

/50/ Stanier,R.Y.: Bact.Rev. 6, 143 - 175 (1942)

/51/ Stanier,R.Y.: J.Bact. 53, 297 - 315 (1947)

/52/ Stewart,J.R., & Brown,R.M.: Arch.Mikrobiol. 80, 176 - 190 (1971)

/53/ Strohl,W.R., & Tait,L.R.: Int.J.Syst.Bact. 28, 293 - 303 (1978)

/54/ Suberkropp,K., & Klug,M.J.: Ecology 57, 707 - 719 (1976)

/55/ Sundarraj,N., & Bhat,J.V.: Arch.Mikrobiol. 85, 159 - 167 (1972)

/56/ Sundarraj,N., & Bhat, J.V.: Arch.Mikrobiol. 77, 155 - 164 (1971)

/57/ Trust,T.J.: J.Appl.Bact. 38, 225 - 233 (1975)

/58/ Turvey,I.R., & Christian,J.: Biochem.J.105, 311 - 315 (1967)

/59/ Tržilova,B., & Miklosovicova,L.: Biologia (Bratislava) 30, 685 - 691 (1975)

/60/ Van der Meulen,A.J., & Harder,W.: Antonie van Leuwenhoek 42, 277 - 286 (1976)

/61/ Veldkamp,H.: J.Gen.Microbiol. 26, 331 -342 (1961)

/62/ Vychodilova,M.: Folia Microbiologica 22, 468 (1977)

/63/ Warke,G.M., & Dhala,S.A.: J.Gen.Microbiol. 51, 43 - 48 (1968)

Acknowledgement: I wish to thank R.R.Colwell,Univ.of Maryland, and R.Y.Morita, Oregon State Univ., for their encouraging support, as well as the German Research Council (DFG) for a scholarship.

# NON-SPREADING CYTOPHAGAS IN ACTIVATED SLUDGE

H. Güde
Landesanstalt für Umweltschutz Baden-Württemberg
Institut für Seenforschung und Fischereiwesen
Untere Seestr. 81, D-7994 Langenargen
Federal Republic of Germany

## Summary

Spreading and non-spreading colonies of yellow pigmented, flexirubin-positive bacteria were observed regularly on agar plates inoculated with samples from different activated sludge and trickling filter sources. One part of the non-spreading colonies formed spreading colonies after transfer to more suitable growth media. The other part was never observed to form spreading colonies on various growth media tested. However, when examined microscopically in wet mount preparations most of the non-spreading isolates showed the same type of gliding motility as the spreading isolates. In order to test the taxonomic homogeneity as well as possible relationships to "classical" cytophagas, 25 non-spreading isolates were examined for 41 morphological and physiological characteristics. Of the strains tested, 19 had more than 80 % of the characteristics in common. Besides their deviating colony growth, the isolates differed from "classical" cytophagas by their more complex growth requirements and by a less wide spectrum of degradation capacities. It is concluded that the non-spreading isolates represent a distinct cluster within a *Cytophaga* group comprising all strains with gliding motility, flexirubin pigmentation and a G + C base ratio in the range of 32 - 36 %.

## Introduction

The widespread occurrence of bacteria belonging to the *Cytophaga* group in marine, freshwater, and soil habitats and the participation of these bacteria in the degradation of natural organic matter is now well documented /3/. Accordingly, one should expect that cytophagas could also represent a component of the microbial biocenoses responsible for biological water purification in sewage plants. Indeed, during a long term study of activated sludge and trickling filter samples, yellow spreading colonies with "classical" Cytophaga characteristics could frequently be detected /5/. Furthermore, considerable numbers of non-spreading yellow pigmented colonies were regularly observed during this study. Of these yellow pigmented colonies, different percentages showed a positive reaction in the KOH test for flexirubin pigmentation, varying with season and sample source. Because of the restricted distribution of flexirubin pigments among eubacteria /17/, the question arose whether these non-spreading, flexirubin positive bacteria might also be related to the Cytophaga group. In the present study it was attempted to answer this question by examining selected isolates of these bacteria from different sewage plants for further morphological and physiological characteristics and by discussing the results in the light of current concepts of Cytophaga taxonomy.

## Materials and Methods

Sampling, plating out, and isolating of the bacteria from the different activated sludge and trickling filter sources have been described in a previous study /5/. The strains further investigated in the present study are listed in Table 1 along with the respective sources they were isolated from.

Table 1: List of the strains investigated and of the respective isolation sources.

| Strain | Source |
|---|---|
| $V_2$, $V_3$, $WB_{29}$, $WB_{32}$, $WB_{34}$ | Sewage plant Wolfenbüttel, activated sludge |
| $SB_2$, $SB_{16}$, $SB_{112}$, $SB_{117}$ | Sewage plant Salzgitter-Lebenstedt, activated sludge |
| $WT_1$, $WT_8$, $WT_4$, $WT_{11}$ | Sewage plant Wolfenbüttel, trickling filter |
| $ST_1$, $ST_2$, $ST_8$, $ST_3$ | Sewage plant Salzgitter-Lebenstedt, trickling filter |
| $SG_8$, $SG_{10}$ | Model activated sludge reactor from "Schering" works, Wolfenbüttel |
| $FN_1$, $FN_2$, $FN_3$ | Sewage plant Friedrichshafen, activated sludge |
| $US_1$, $US_7$, $US_8$ | Sewage plant Unteres Schussental, activated sludge |
| $C_R$ | Lake Mindel (near Lake Constance) |

Tests for spreading growth and for gliding motility were performed on nutrient agar (NA, Merck), Casitone-yeast extract (Cy) agar, Vy/2 agar (yeast cells + vitamin $B_{12}$), and activated sludge extract (ASE) agar. The agar media were prepared as described before /5/. Microscopical examination of gliding motility was done with cells grown for 24 h on Vy/2 or ASE- agar. The bacteria were observed under the microscope after transfer of cell material from the edge of the colony together with a small piece of the underlying agar to the microscopical slide. Flagellation was tested with Leifson's stain /6/.

Anaerobic growth was tested on nutrient agar in anaerobic jars incubated for 2 weeks at 25°C. Growth temperature studies were performed with bacteria growing in shaking flasks on nutrient broth (NB, Merck) at the respective temperatures. Growth requirements of the bacteria were studied in liquid media containing a mineral salts medium and either glucose (5 g/l), or glucose + Bacto casamino acids (5 g/l, Difco), or glucose + casamino acids + a vitamin solution.

All biochemical tests were carried out according to the prescriptions given in the handbook of Harrigan and McCance /6/.

## Results

Because the formation of spreading colonies can depend on the composition of the agar media used /3/, it was suggested that the flexirubin positive bacteria not spreading on the original enumeration and isolation agar (Cy agar) would perhaps demonstrate spreading growth on other agar media. Indeed, 22 out of 59 strains not spreading on the nutrient rich NA and Cy media were shown to form spreading colonies on ASE or Vy/2 agar which are comparatively poor in easily available substrates (Table 2). Consequently, these spreading strains could be regarded as fitting into the "classical" Cytophaga concept and were not further investigated here. Therefore, in the following the investigation has been concentrated on those strains which could never be shown to form spreading colonies in spite of the fact that gliding motility was observed with the majority of them (Table 2).

Table 2: Comparison of spreading growth and gliding motility on different agar media among 59 flexirubin-positive isolates from originally non-spreading colonies.

| Agar media | NA | Cy | ASE | Vy/2 |
|---|---|---|---|---|
| Spreading growth | 0 | 0 | 22 | 23 |
| Gliding motility | 9 | 15 | 51 | 51 |

Among 25 of these non-spreading strains, 19 were microscopically observed to glide (Table 3). These were put together in group A. Group B consisted of the remaining 6 strains (strains $WT_1$, $WT_{11}$, $ST_3$, $SG_8$, $SG_{10}$, and $US_8$). Both strains were compared with strain $C_R$ which could be considered as a typical Cytophaga sp.

Table 3: Comparison of morphological and growth characteristics of flexirubin-positive, non-spreading strains with those of the Cytophaga strain $C_R$. Group A consisted of 19 strains, group B of 6 strains. Numbers represent percentage frequencies of positive characters among the respective strains.

| Characteristics | Group A | Group B | Strain $C_R$ |
|---|---|---|---|
| Orange-yellow pigmentation on NA | 100 | 66 | + |
| Spreading on Vy/2 agar | 0 | 0 | + |
| Gliding motility | 100 | 0 | + |
| Flagellation | 0 | 0 | - |
| Short ( 3 $\mu$m) rods on ASE agar | 100 | 100 | + |
| Elongated rods ( 8 $\mu$m) on Cy agar | 89 | 33 | + |
| Anaerobic growth | 11 | 50 | ± |
| Growth in NB at | | | |
| 5°C | 74 | 0 | + |
| 10°C | 95 | 33 | + |
| 20°C | 100 | 100 | + |
| 30°C | 79 | 100 | - |
| 37°C | 5 | 17 | - |
| Growth in mineral media + glucose | 0 | 17 | + |
| Growth in mineral media + glucose + amino acids | 68 | 66 | + |
| Growth in mineral media + glucose + amino acids + vitamins | 95 | 100 | + |

Morphologically, it was common to group A and strain $C_R$ that the cells grew as elongated rods in nutrient rich media, whereas only small rods were observed in nutrient poor media (Table 3). In contrast the majority of group B grew as small rods in both media.

Table 4: Comparison of biochemical characteristics of group A, group B and strain $C_R$ (further explications: see Table 3)

| Characteristics | Group A | Group B | Strain $C_R$ |
|---|---|---|---|
| Utilization of: | | | |
| Glucose | 83 | 66 | + |
| Galactose | 68 | 50 | + |
| Mannose | 63 | 50 | + |
| Lactose | 58 | 50 | + |
| Maltose | 63 | 50 | + |
| Ribose | 42 | 33 | + |
| Arabinose | 74 | 66 | + |
| Xylose | 79 | 66 | + |
| Citrate | 5 | 33 | - |
| Lysis of: | | | |
| Casein | 100 | 83 | + |
| Blood | 83 | 66 | - |
| Gelatin | 89 | 66 | + |
| Chitin | 0 | 0 | + |
| Dextran | 0 | 0 | + |
| Pectin | 0 | 0 | + |
| Starch | 68 | 33 | + |
| Xylan | 0 | 0 | + |
| Yeast cells | 68 | 50 | + |
| Catalase | 95 | 33 | + |
| Oxidase | 89 | 33 | + |
| Phosphatase | 100 | 66 | + |
| Lecithinase | 95 | 33 | - |
| Urease | 0 | 50 | - |
| Nitrate reduction | 63 | 50 | + |
| $H_2S$ formation | 0 | 33 | - |
| Indole formation | 11 | 17 | - |

With respect to growth characteristics, there were remarkable differences between group A and strain $C_R$ (Table 3). The latter showed better growth at lower temperatures and did not grow at temperatures of 30 $^{\circ}C$ and above. Furthermore, strain $C_R$ had many fewer growth requirements than the majority of group A strains. Group B strains also grew better at higher temperatures and had more complex growth requirements. Additionally, 50 % of them could grow well anaerobically, a characteristic only seldom observed in group A strains, and only weakly expressed in strain $C_R$.

Also with regard to the biochemical characteristics there were considerable differences between group A and strain $C_R$ (Table 4). No member of group A strains was able to attack any of the polysaccharides except starch. Additionally, the utilization of sugars was more restricted among the group A strains, whereas haemolysis and lecithinase was not observed with strain $C_R$. However, it must not be overlooked that there were also several characteristics in common such as the presence of catalase, oxidase, and phosphatase, and the absence of urease, $H_2S$ and indole formation, and the lack of citrate utilization. Also in this case, significant differences between group A and group B could be observed. Furthermore, group A with a mean similarity of 87 % (calculated on the base of negative and positive matches) proved to be much more homogenous than group B, the corresponding similarity value of which amounted to only 67 %.

## Discussion

The non-spreading activated sludge isolates investigated in this study demonstrated a lot of similarities, but also several important differences to typical cytophagas. Therefore, the question as to the relationship of these bacteria to the Cytophaga group can not be answered without entering deeper into the problems concerning the taxonomic differentiation of this group.

Because of the lacking demonstration of spreading growth in all strains and because of the failing of gliding motility in group B strains, it becomes obvious that also in this case the long discussed problem of the differentiation between cytophagas and flavobacteria proves to be of central importance /3, 7, 9, 10, 12, 13, 17, 19, 20/. At present, the most accepted strategy is to allocate the non-motile strains with a low G + C content to the Flavobacterium group, whereas the Cytophaga group would comprise the low G + C base content bacteria showing unequivocal gliding motility. Accordingly, at least group A strains could be regarded as cytophagas due to a low G + C base content (determined for two strains by Behrens /1/) and due to gliding motility. However, it must be mentioned that the type of motion observed with group A strains resembled to a high degree that described by Perry /15/ inasmuch as the bacteria glided only over comparatively short distances (maximum three cell lengths) with frequent changes of direction. It has been argued /7,12/ that Perry's observations would perhaps be better described as what has been defined by Henrichsen /8/ as "twitching" motion. However, Acinetobacter calcoaceticus, a typical representative of twitching strains showed jerky and irregular motions, whereas the movements of group A strains regularly followed the long axis of the cell. Additionally, typical cytophagas also moved in the same way when observed under the described conditions.

In any case, it becomes apparent by this example that a taxonomic differentiation based on gliding motility alone, must be regarded as unsatisfactory. Thus, the relationships between Cytophaga and Flexibacter were never doubted, in spite of the fact the G + C base contents of this group showed remarkable differences /1/. In contrast, relationships between cytophagas and flavobacteria would implicitly be excluded by the application of this criterium, although they could be much more frequent because of similar G + C content values /1/.

These considerations make it evident that final decisions as to these relationships can only be made on the base of genetic homology studies with these strains. A first step in this direction was done by Behrens /1/ who investigated the DNA homology of strains having a G + C base content in the range of 32 - 36 %, including two strains of group A ($V_3$ and $SB_2$). Both strains were shown to be closely related to each other, whereas homologies to other cytophagas were much lower. However, it must be noted that also many other cytophagas which phenotypically would clearly have been designated as typical, were demonstrated to have low DNA homology values.

After all, it can be stated that the group A strains at least can be considered as being related to the Cytophaga group, based on DNA characteristics, pigmentation patterns (studied with two isolates by Fautz /4/), and on the morphological and physiological characteristics (investigated in this study and by Hirsch /9/). Thus, these bacteria would represent a distinct cluster within a Cytophaga group comprising several clusters with the common characteristics of gliding motility, flexirubin pigmentation, and a G + C base content in the range of 32 - 34 %.

Here, it can not be decided whether the occurrence of these bacteria is restricted to sewage plant habitats, because so far no strains from other sources were examined. However, the fact that these bacteria could easily be found in several different sewage plants indicate that they may represent regular inhabitants of activated sludge and trickling filter biocenoses. Consequently, it can be assumed that at least a part of those bacteria designated as flavobacteria may belong to this Cytophaga-cluster /1, 11, 16/.

## References


/1/ Behrens, H.: Dissertation. TU Braunschweig (1978)

/2/ Benedict, R. G., Carlson, D. A.: Water Res. 5, 1023 - 1030 (1971)

/3) Christensen, P.: Can. J. Microbiol. 23, 1599 - 1653 (1977)

/4/ Fautz, E. R.: Dissertation. TU Braunschweig (1978)

/5/ Güde, H.: Appl. Environm. Microbiol. 39, 756 - 763 (1980)

/6/ Harrigan, W. F., McCance, M.E.: Laboratory methods in microbiology London (1966) Academic Press.

/7/ Hayes, P. R.: J. Appl. Bacteriol. 43, 345 - 367 (1977)

/8/ Henrichsen, J.: Bacteriol. Rev. 36, 478 - 503 (1972)

/9/ Hirsch, I.: Dissertation. TU Braunschweig (1980)

/10/ Holmes, B., Owen, R. J.: Int. J. Syst. Bacteriol. 29, 416 - 426 (1979)

/11/ Lighthart, B., Oglesby, R. T.: J. Water Pollut. Control Fed. 41, R 267 - 281 (1969)

/12/ McMeekin, T. A., Stewart, D. B., Murray, J. G.: J. Appl. Bacteriol. 35, 129 - 137 (1972)

/13/ McMeekin, T. G., Shewan, J. M.: J. Appl. Bacteriol. 45, 321 - 332 (1978)

/14/ Mitchell, T. G., Shewan, J. M., Hendrie, M. S.: J. Appl. Bacteriol. 32, 40 - 50 (1969)

/15/ Perry, L. B.: J. Appl. Bacteriol. 36, 227 - 232 (1973)

/16/ Pike, L. F.: p. 1 - 63 in : Curds, C. R., Hawkes, H. A. (eds.) Ecological aspects of used water treatment. London (1975) Academic Press

/17/ Reichardt, W.: Zbl. Bakteriol. Parasitenkd. Infektionskr.Hyg. Abt. 1 Orig. 227, 85 - 93 (1974)

/18/ Reichenbach, H. Kleinig, H. Achenbach, H.: Arch. Microbiol. 101, 131 - 144 (1974)

/19/ Weeks, O. B.: J. Appl. Bacteriol. 32, 13 - 18 (1969)

/20/ Weeks, O. B.: p. 357 - 364 in : Buchanan, R. E., Gibbons, N.E. (eds.) Bergey's manual of determinative bacteriology Baltimore (1974) Williams and Wilkins

# STUDIES ON THE DEGRADATION OF CELLULOSE BY SPOROCYTOPHAGA MYXOCOCCOIDES

Rosemary M. Griffiths, J.H. Parish
University of Leeds, Department of Biochemistry, 9 Hyde Terrace,
Leeds LS2 9LS, U.K.

## Summary

Culture fluids of *Sporocytophaga myxococcoides* contain extracellular enzyme activities capable of hydrolysing cellulose, carboxymethylcellulose and cellobiose. The yield of enzyme activities varies with the nature of the organism's carbon source and glucose is apparently an inducer. There is no detectable glucanase capable of hydrolysing β-nitrophenyl glucoside in these preparations. Supernatants of cellulose grown cells do not contain any detectable glucose or cellobiose. Such supernatants contain, in addition to cellulose, an unidentified reducing compound and no other detectable non-ionic carbon compounds. The reducing compound functions as a linking metabolite in the growth of a non-cellulolytic *Alcaligenes* sp. that is capable of growth in mixed culture with *S. myxococcoides* using cellulose as sole source of carbon and energy.

## Introduction

*Sporocytophaga myxococcoides* is a cellulose degrading,microcyst forming cytophaga, usually described as requiring glucose or cellulose for growth /5/. The organism is of considerable scientific and potential biotechnological interest. There is general agreement that the organism produces an extracellular cellulase complex and this seems to resemble that of certain fungi (*Trichoderma reesei*, *T. koningii* and related organisms) rather than that of other cellulolytic bacteria /3/.

Assay of cell-free culture supernatants reveals carboxymethylcellulase activity that can be detected either viscometrically or by release of reducing sugars. The activity consists of a mixture of enzymes, at least two of which are endoglucanases /9/. It has been reported that these extracellular enzymes do not include an activity for the hydrolysis of native celluloses /1,9/ and this has been correlated with the finding that the organism produces cavities in its substrates suggestive of the activity of a cell-bound cellulase /1/. However, other workers report the presence in the supernatants of an activity that hydrolyses very finely divided cellulose (Avicel) and they ascribe the difference to the fact that, unlike the previous reports, they employed an assay for soluble carbohydrate rather than reducing sugars and this assay is supposed to be more sensitive /12/.

We have re-examined the extracellular cellulases of S. myxococcoides and have studied the production of non-cellulosic compounds by cultures of the organism *in vivo*.

## Materials and Methods

Organisms and culture conditions: S. myxococcoides NCIB8639 and a bacterium, strain 10G (tentatively assigned to Alcaligenes) were employed for these studies. Stanier's nitrate medium /4/ was employed throughout. Sugars were added to 1% unless stated to the contrary. Two cellulosic substrates, Whatman paper powder CC41 and bacterial cellulose mats /8/, were employed.

Biochemical and chemical methods: Culture supernatants were concentrated by freeze-drying and re-dissolving. For enzyme assays the solutions were dialysed and the final buffer concentration was 1 mM Tris-HCl pH 7.5. Glucose was assayed by both the glucose oxidase/peroxidase method and with hexokinase, glucose 6-phosphate dehydrogenase, ATP and $NAD^+$. Total carbohydrate was measured by the anthrone/sulphuric acid method /10/ and total cellulose was measured by an adaptation of this method for alkali-insoluble material. Reducing sugars were measured by a quantitative variation of the tetrazolium procedure /2/; the insoluble blue formazan produced in this reaction was extracted with toluene and the absorption of the solution was measured at 570 nm. In both of these cases, the reagents were calibrated with standard glucose solutions and the results are presented as glucose equivalents. Sugars in culture supernatants were assayed by thin-layer chromatography /7/. Estimates of lower limits of cellobiose were obtained by noting the amount of the sugar that was just detectable in chromatograms of standards. Protein was assayed by using the Lowry reagent calibrated with bovine serum albumin.

Measurement of growth: Growth in media containing sugars was monitored turbimetrically; growth of S. myxococcoides in cellulose was monitored by counting cells with a haemocytometer; growth of Alcaligenes 10G in mixed cultures was monitored by dilution and plate counts.

## Results

Glucose, cellobiose, cellulose and mannose supported growth of S. myxococcoides; carboxymethylcellulose, starch, maltose, lactose, arabinose, galactose, fructose, mannitol, pyruvate, lactate, citrate, succinate and glutamate did not. Cellulose was degraded in the presence of glucose. The organism was found to grow at comparable rates in 1% glucose and 1% cellulose; it also grows in mannose but significantly more slowly (Table 1). The growth on glucose is sensitive to substrate concentration; that on cellulose is not. By plotting the reciprocal of specific growth rate as a function of the reciprocal of substrate concentration over the range 0.1-2.0% substrate concentrations (data not shown) we calculated the apparent $K_m$ values for these two substrates. The organism has a high apparent $K_m$ (low affinity) for glucose. We believe that this reflects a property of the glucose permease system because the intracellular hexokinase of *S. myxococcoides* has a very low $K_m$ (R. Mullings, unpublished data). As there is no detectable glucose in the supernatants of cellulose grown cells (Table 1), we conclude that the cells transport a product of cellulose degradation other than glucose when the cells are grown in cellulose.

Table 1. Summary of the utilization of carbon sources by *S. myxococcoides*

| | | |
|---|---|---|
| Mean doubling times in ... | 1% glucose | 4.25 h |
| | 1% cellulose | 4.0 h |
| | 1% mannose | 7.5 h |
| | 0.2% glucose | 11.0 h |
| | 0.2% cellulose | 4.0 h |
| Apparent $K_m$ for growth at expense of glucose is | | 0.3 % |
| Apparent $K_m$ for growth at expense of cellulose is | | <0.05 % |
| Concentration of glucose accumulated in cellulose cultures is | | <15 µg/100 ml |
| Concentration of cellobiose accumulated in cellulose cultures is | | <500 µg/100 ml |

Extracellular enzymes were assayed qualitatively in supernatants of cultures. A cellulase activity was detected that was capable of forming a hole in a bacterial cellulose mat. We were unable to detect a conventional β(1-4) glucanase capable of hydrolysing an artificial nitrophenyl glucoside substrate at any pH between 4 and 10. We were also unable to detect any activity capable of releasing soluble chromophore from cellulose dyed with remazol brilliant blue R /7/.

Quantitative analysis of enzymes in supernatants revealed cellulase, carboxymethyl cellulase and cellobiase. The specific activities of these enzymes suggested that their production was dependent to some extent upon the carbon source and that, in particular, glucose appears to act as an inducer (Table 2). The cellobiase activity in the table appears to be that of a genuine cellobiose hydrolase and not a cellobiose phosphorylase. Phosphate is not required for the reaction and in experiments (not shown) in which low concentrations of inorganic phosphate were intentionally included in the incubation medium, there was no detectable conversion of inorganic phosphate to sugar phosphate.

Table 2. Specific activities of cellulase (using CC41 paper powder as substrate), carboxymethylcellulase (CMCase) and cellobiase in culture supernatants of S. myxococcoides. One unit is defined as the rate of increase in glucose (for cellobiase) or glucose reducing equivalents (for cellulase and CMCase) in mg/ml·h. Entries in the table are specific activities in units/mg protein x 1000.

| Growth substrate | Cellulase | CMCase | Cellobiase |
|---|---|---|---|
| Cellobiose | 3.3 | 19.5 | 68 |
| Glucose | 17.4 | 24.8 | 263 |
| Mannose | 8.6 | 11.7 | 21 |
| Cellulose | 9.3 | 3.4 | 3 |

Cultures of S. myxococcoides growing in cellulose accumulate a reducing compound. The chemical identity of this substance is unknown. It reacts with anthrone, although less strongly than glucose (the anthrone/reducing equivalents ratio is approximately one half of the equivalent value for glucose). It reduces the tetrazolium reagent and also $Ag^+/NH^+{}_4$. Solutions of the compound have an absorption maximum at 290 nm, suggestive of an unconjugated free aldehyde group. It does not produce a spot corresponding to any common aldose on thin-layer chromatograms. Alcaligenes sp. 10G grows on glucose and cellobiose but not cellulose. However it will grow in association with S. myxococcoides on cellulose and will also grow on the reducing material isolated from supernatants of S. myxococcoides cultures grown on cellulose.

The growth of these organisms in cellulose is summarized in Fig. 1. The reducing compound continued to accumulate after the organisms had reached stationary phase. In mixed cultures the accumulation of the reducing compound was inhibited and the Alcaligenes grew apparently at the expense of this compound because the yield of S. myxococcoides in the mixed culture was not significantly different from the yield in pure culture. The limited growth of Alcaligenes in pure culture in cellulose medium was presumably due to contaminating sugars (or other available carbon sources) in the paper powder. There was no detectable change in the apparent chain length of residual cellulose during growth. The apparent cellulose chain length (approximately 100) obtained by measuring the ratio of anthrone equivalents to reducing equivalents is an underestimate by an order of magnitude; this is because the colour yield by a glucose residue in cellulose is less than that of free glucose. However, the data do confirm that no detectable short-chain cellulose accumulates in the cultures.

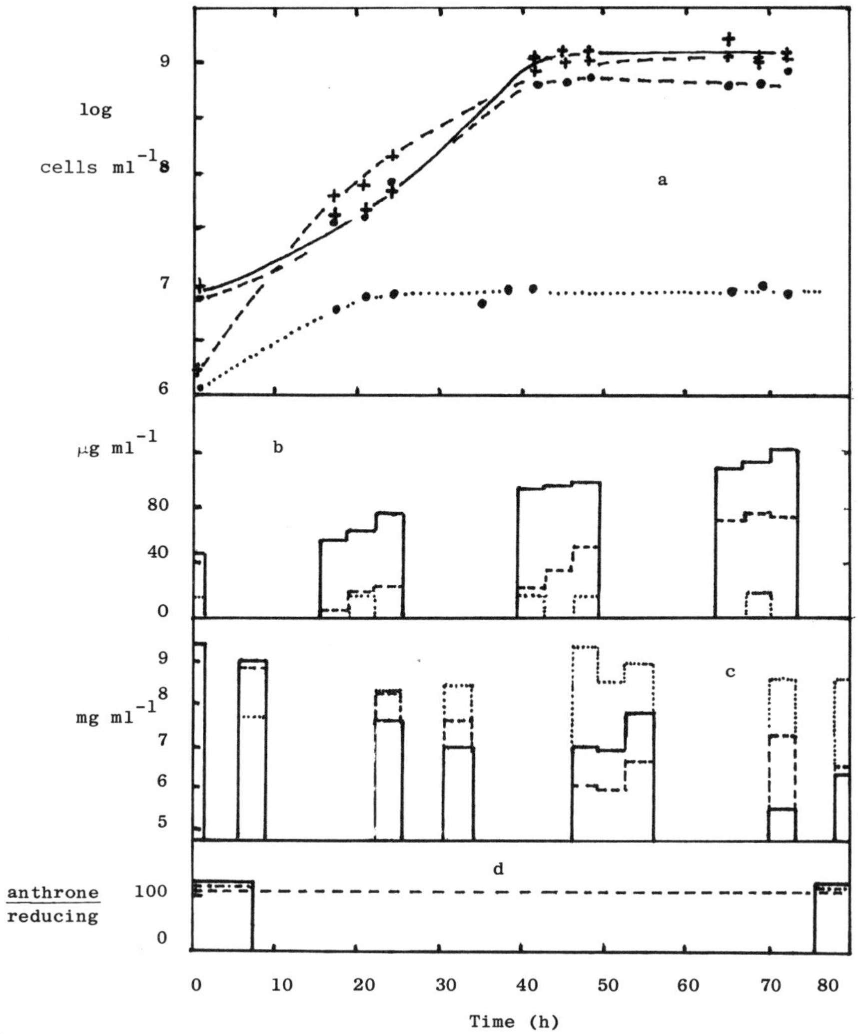


Fig. 1. Growth of cells in cellulose. Throughout the figure, continuous lines refer to S. myxococcoides in pure culture, dotted lines to Alcaligenes sp. 10G in pure culture and discontinuous (dashed) lines to a mixed culture. (a) Growth of cells, + S. myxococcoides, • Alcaligenes. (b) Reducing compound(s) in the supernatants in glucose equivalents. (c) Residual cellulose in glucose-anthrone equivalents. (d) Ratio of anthrone equivalents to reducing equivalents for the residual cellulose.

## Discussion

Our data confirm that cellulase, as defined as an enzyme complex capable of hydrolysing paper cellulose is present in culture supernatants of S. myxococcoides even when paper powder (regarded as a "difficult" substrate /12/ is employed. It is important to employ a suitably sensitive assay for reducing sugar to detect the enzyme.

The enzyme activities assayed in these supernatants (Table 1) are presumed to be due to components of the cellulase complex. Carboxymethylcellulase is a convenient activity for assay but does not reflect the use of a physiological substrate because the organism will not grow on this material. Likewise we believe the cellobiase may have substrates other than cellobiose because it is produced in cellulose grown cells and such cultures do not contain detectable cellobiose. The activities are dependent upon the nature of carbon source. Previously it had been shown that the production of extracellular enzymes is dependent on the nature of the cellulose employed /12/. The survey of every one of the extremely limited number of carbon sources available for this organism suggests that glucose is an inducer (Table 2). This finding runs counter to metabolic regulation in bacteria in which glucose typically represses production of degradative enzymes through the catabolite effect. The explanation may lie in a contaminant of the glucose. The remarkably interesting disaccharide sophorose, β(1-2) glucosyl glucose, is present in small amounts in all commercial glucose. It is apparently produced as a byproduct of the mid hydrolysis of starch by a rearrangement. Sophorose is an inducer of cellulose in Trichoderma /11/ and it is possible that it also acts as a gratuitous inducer of cellobiase and other components of the cellulose complex in S. myxococcoides.

We postulate that certain products of cellulose degradation, presumptively a group of cellulose oligosaccharides, are taken up by the cells by a very affinity permease system with the result that these compounds do not accumulate in appreciable concentrations in the medium. The reducing compound, that cross-feeds Alcaligenes sp. 10G, is apparently a secondary metabolite of S. myxococcoides and growth of the organism is indifferent to accumulation of the compound.

## Acknowledgements

We are grateful to Dr H.A. Painter of the Water Research Centre, Stevenage for discussions, advice and access to facilities and to Mr A. Paskins for the gift of Alcaligenes sp. 10G. R.M.G. is supported by an SRC CASE Research Studentship.

## References

/1/ Berg, B., v. Hofsten, B., Pettersson, G.: J. appl. Bacteriol. 35, 215-219 (1972)

/2/ Cheronis, N.D.: Mikrochim. Acta, part 1, 924-947 (1956)

/3/ Gooday, G.W.: p. 437-460 in: Berkeley, R.C.S., Gooday, G.W., Ellwood, D.C. (eds.), Microbial polysaccharides and polysaccharases. London (1979): Academic Press

/4/ Lampky, J.R.: J. Bacteriol. 126, 1278-1284 (1976)

/5/ Leadbetter, E.R.: p. 101-112 in: Buchanan, R.E., Gibbons, N.E. (eds.), Bergey's Mannual of Determinative Bacteriology, 8th edition. Baltimore (1974): Williams & Wilkins

/6/ Leisola, M., Linko, M., Karvonen, E.: p. 297-313 in: Bailey, M., Enar, T.M., Linko, M. (eds.), Symposium on enzymatic hydrolysis of cellulose. Helsinki (1975): Finnish National Fund for Research and Development

/7/ Lewis, B.A., Smith, F.: p. 807-837 in: Stahl, E. (ed.), Thin-layer Chromatography. Berlin (1969): Springer-Verlag

/8/ Mullings, R., Parish, J.H.: this volume

/9/ Osmundsvag, K., Goksøyr, J.: Eur. J. Biochem. 57, 405-409 (1975)

/10/ Spiro, R.G.: Meth. Enzymol. 8, 4-5 (1966)

/11/ Sternberg, D., Mandels, G.R.: J. Bacteriol. 139, 761-769 (1979)

/12/ Vance, I., Topham, C.M., Blaydon, S.L., Tampion, J.: J. gen. Microbiol. 117, 235-241 (1980)

RESULTS OF A DISCUSSION ABOUT TECHNICAL PROBLEMS IN TAXONOMIC RESEARCH ON FLAVOBACTERIA AND CYTOPHAGA-LIKE BACTERIA

The participants of the symposium felt, that an exchange of results and ideas between the two groups of investigators working with flavobacteria on the one side, and with Cytophaga-like bacteria on the other, could potentially be very useful and stimulating for both sides, particularly as the borders between some of these bacteria tend to vanish. Closer contacts within each group of microbiologists, and between the two, would be highly desirable.

To promote such contacts it was proposed, a) to start a Newsletter on Flavobacteria and Cytophaga-Like Bacteria, and b) to repeat meetings of the kind just concluded from time to time. Dr. O.B. Weeks from Las Cruces, New Mexico, kindly took over responsibility for compiling and distributing the Newsletter.*

A serious problem experienced by all taxonomists is the availability of reference specimens. Today type strains of all valid taxa of microorganisms are required to be deposited at a culture collection, whenever this is feasible, and are thus generally accessible. In contrast, interesting and even important strains, which appear in the course of a taxonomic study but represent only unnamed or preliminary phena or taxa, are in constant danger of getting lost, and are often no longer available even a short time after publication of the results of the study. It was therefore proposed to start temporary collections which are to preserve all organisms representing certain types, phena, or clusters during the period in which the classification of the respective bacteria is being worked out. This would guarantee that future investigators could continue at exactly the point where an earlier investigator has stopped. Drs. Holmes and Owen promised to inquire whether the collection of flavobacteria could be kept at NCTC in London, while Dr. Claus agreed to take care of the collection of Cytophaga-like bacteria at the DSM (German Collection of Microorganisms) in Göttingen.

H. R.

*The untimely and sudden death of Dr. Weeks prevented him from organizing the Newsletter. There was, however, a note from him in the January 1981 issue of ASM News announcing the newsletter. Microbiologists from many different places had already signaled their interest, and quick action was mandatory. We are indebted to Dr. R.J. Owen that he agreed to step in and to take care of the Newsletter, at least for the time being. Dr. Owen is in an excellent position to handle the Newsletter, as he is a well-known specialist for flavobacteria, will be co-author of the Flavobacterium section in the new Bergey edition, and his laboratory at NCTC will probably be responsible for the planned reference collection of flavobacteria. Persons interested in contributing to, and in obtaining the Newsletter should contact:

Dr. R.J. Owen
National Collection of Type Cultures
Central Public Health Laboratory
Colindale Avenue
London NW9 5HT England.

H. R.